MICHEL ODENT, MD, was in charge of the surgical and maternity units at the Pithiviers state hospital in France during the years 1962-1985. For many years he was the only doctor in charge of around 1,000 yearly births. He is the author of the first article in medical literature about the use of birthing pools (*Lancet*, 1983), the first article about the initiation of lactation during the hour following birth (1977), and the first article applying the 'Gate Control Theory of Pain' to obstetrics (1975). He coined the term 'hormone of love' when mentioning oxytocin. He created the Primal Health Research database www.primalhealthresearch.com, and has been a member of the Professional Advisory Board of La Leche League International for some 40 years. Michel Odent is Visiting Professor at the Odessa National Medical University and Doctor Honoris Causa of the University of Brasilia.

PLANET OCEAN

Our Mysterious Connections to Water

MICHEL ODENT

Clairview Books Ltd.
Russet, Sandy Lane,
West Hoathly,
W. Sussex RH19 4QQ

www.clairviewbooks.com

Published by Clairview Books 2021

A CIP catalogue record for this book is available from the British Library

ISBN 978 1 912992 27 0

Cover by Morgan Creative
Typeset by Symbiosys Technologies, Vishakapatnam, India
Printed and bound by 4Edge Ltd, Essex

Contents

Abstracts

1. Mysteries

The objective of this preliminary chapter is to provide reasons to identify and interpret mysterious human characteristics. We refer to modern life and also to prehistory, history, myths, legends and arts as diverse as poetry, sculpture, painting, songs and cinema. We wonder why the small groups of scientists who raised questions about the dozens of traits humans share with sea mammals, but not with other primates, have been marginalized.

2. Fluctuations of sea levels

Measurable criteria
Palaeomicrobiology
The prototype of a modern interdisciplinary scholar

We present the evolution of our planet and the evolution of Homo as two inseparable topics. That is why we find it necessary to recall, in particular, that during a period covering the last two million years sea levels have fluctuated some 218 metres.

An estimated 20 million km² of territory was exposed on the world's continental shelves between 110,000 and 10,000 years before the present. Exploitation of marine resources has undoubtedly created the potential for relatively high human population densities. During this period, some of the major phases of human history have taken place, including renewed human dispersals out of Africa.

Such facts suggest that until now paleoanthropologists might have studied marginalized human populations that colonized inland territories, probably after migrations along rivers and around lakes.

3. Homo navigator

The Mediterranean Basin
Questions from Gibraltar
The Pacific Rim
From Madagascar to Easter Island

There are countless published documents about the palaeolithic hunter-gatherers. The time has come to realize that 'Planet Ocean' has been to a great extent colonized by 'Homo Navigator'. We focus on the Mediterranean Basin and the Pacific Rim.

By raising questions about this little-known variety of prehistoric Homo, we take another look at human nature.

4. Looking towards the future?

Limited perspectives
A unique dominant status
Why the Americas before Madagascar?
Underwater fossil hunting

As interdisciplinary students in human nature, we are combining records of the past in order to look towards the future.

Each perspective is in the hand of specialized researchers and has limited objectives.

Fossil hunting (palaeoanthropology) has narrow limits: when human fossils are found it means first that the geological and environmental conditions are absolutely exceptional. The advent of valuable fossil-dating methods has enlarged its power.

Archaeology, which looks at physical objects shaped by human contact, also has obvious narrow limits.

Today human evolutionary genetics has reached a unique dominant status. It is established that since the glacial maximum, a land-mass the size of South America has been inundated worldwide. On the day when this piece of information becomes common knowledge, underwater fossil hunting (i.e., underwater palaeoanthropology) might become an irreplaceable complementary key to studying human nature.

5. Homo's best friends

The dolphin connection
The canine connection

How to interpret the deep-rooted mysterious connections between Homo and dolphins, on the one hand, and Homo and dogs, on the other hand.

6. The super brainy mammals: Homo and Tucuxi

Nutritional needs
The dolphin family
Too rational to survive

Questions about common points, advantages and handicaps of mammals endowed with a supersized brain inspire joint studies of Homo and members of the dolphin family. The focus should be on nutritional needs and enzymatic systems.

7. From Aesop to Elaine Morgan: What pioneers have in common

The pioneers who focused on various human characteristics suggestive of an aquatic past had several common points. One of them leads to observe that interdisciplinarity, as the capacity to focus on links between seemingly different areas, is a facet of lateral thinking.

8. Straight to the point

It should be enough
It is not enough
One step further

The defining feature of Homo is a supersized brain. From this perspective, human beings have more similarities with sea mammals than with land mammals in general and the other apes in particular. As a general rule, when a trait is mysterious, and apparently specific to humans, there are reasons to look at what we have in common with mammals adapted to the sea.

In the age of genetics, we suggest that the best way to challenge the dominant theories and to go one step further is to restart from the central event that separated the emerging *Homo* from *Pan* (the chimpanzees). We now have reasons to suggest that this event is at the root of the defining feature of Homo.

9. The mammal that cannot swim

Learning from babies
Learning from non-human apes
From Plato to the next Olympic Games

Breaststroke versus front crawl: the realm of mysteries
Advantages of learned behaviours

Humans are the mammals that need to learn a technique to be able to swim. They lose their innate capacity to swim when they reach a certain degree of neocortical maturity that takes place around the age of three or four months.

The great diversity of styles confirms that, beyond a certain degree of brain development, swimming must be classified among learned behaviours: it occurs only after experience or practice. A learned behaviour is an advantage in terms of adaptability: it is flexible.

10. Why is human birth occasionally easy?

When the tool becomes the master
The solution found by nature
The foetus ejection reflex
Challenging thousands of years of tradition

Until now, studies of birth physiology among humans were based on interpretations of difficulties. This is why the comparative importance of mechanical factors has been overestimated by theoreticians. We should first interpret the well-known fact that, occasionally, women who are not special from a morphological perspective can give birth easily and quickly, while others need medical intervention after days of hard labour. This enormous discrepancy leads us to understand that birth physiology is first and foremost a chapter of brain physiology. It is essential to realize that the part of the brain that has reached an extremely high level of development in our species—the 'new brain' or neocortex—does not always play the role of a tool at the service of vital physiological functions. On the contrary, in some particular situations, it can inhibit and weaken such functions. It is as if the tool may become the master. The main objective of this chapter is to popularize the concept of neocortical inhibition. Reduced self-control, as an effect of a reduced neocortical activity, appears as the main factor that makes human birth possible. When this solution found by nature is understood, it becomes easy to analyze and summarize the basic needs of a labouring woman: she needs to feel protected against all possible neocortical stimulations. The keyword is protection. The main stimulants of neocortical activity are well-known: language, light and all attention-enhancing situations.

11. The Oceanic feeling

The highways to transcendence
Half-open doors to transcendence

During transcendent emotional states, human beings can escape from space and time reality. Episodes of human reproductive life such as the 'foetus ejection reflex', the 'milk ejection reflex' and the 'sperm ejection reflex' are presented as 'the highways to transcendence'. Enthusiasm and joy are considered in this framework.

12. Humanity and Mother Ocean

From Green to Blue
Disclosing priorities
Beyond the explicit messages

Modern humans must urgently realize that 95 per cent of the planet's habitable space lies within the deep immense oceans: the oceans may be presented as the givers and sustainers of life. In such a context we rephrase a common question: how can humanity develop its respect for 'Mother Ocean'?

13. From a garden paddling pool to the Pacific Ocean

Jumping from one scale to another
In the age of tap water
Before the age of tap water

Homo—the primate endowed with a powerful neocortical supercomputer—has an immense capacity to jump from one order of magnitude to another. I illustrate this capacity in the framework of the relationship between Homo and water.

<u>1</u>

Mysteries

In 1934, while my family was living in a village in Northern France, we got a car. The following Sunday, we went to the seashore by the Channel. A dream had come true!

In 1936, a paid vacation policy was set by a new French government. Millions of people went to the beach and spent days watching the waves, gazing at the horizon … and dreaming. This was an inspiration for my mother as a poet. The first verse of her poem in alexandrine titled '*Vers ton île*' (Towards Your Island) was a question:

> '*Qui donc m'emportera vers ton île lointaine?*'
> (Who will take me to your distant island?)

The French navigator Jacques Cousteau was curious about his own mysterious vocation, confessing that he did not know why he loved the sea.

Once I was in a shop in Tokyo. There was a musical background. We could hear '*La Mer*', by Charles Trenet. A poemsong about the sea makes people dream all over the world.

Myths and legends also belong to the realm of irrationality The mermaid legends are mysterious because they are universal. There is always an erotic aura surrounding them. The typical mermaid has the head and body of a woman. Her hair is long and beautiful. Below the waist the mermaid is like a fish. The mermaid represents the femme fatale of the sea. Is there a link between the mermaid legends and the erotic power of dresses that do not separate the legs of women? Many other legends are based on the close relationship between the mysterious Eros and the sea.[1] It is significant that Aphrodite, the Goddess of Love, was born from the foam of the waves.

In every land, in every age, poets have testified to the irresistible erotic power of the sea. Let us recall 'Song for St Cecilia's Day', the poem by W.H. Auden:

> Blonde Aphrodite rose up excited,
> Moved to delight by the melody,
> White as an orchid she rode quite naked
> In an oyster shell on top of the sea; …

It would take volumes to present an overview of works of art inspired by mermaids, not just the famous sculpture in Copenhagen.

The mermaid comes up again and again in music. In the catalogue of modern music in the British Library the list of works associated with the key word 'mermaid' takes up many pages.

One could re-examine the entire history of painting from the angle of the relationship between Homo and the sea and, in particular, the erotic power of the sea. The collection by Genie Kahler titled 'Women and the Sea' leads to hundreds of pins!

We must even consider the case of cinema as an artistic medium that could not be used before the twentieth century. In films, the sea has frequently been used as a powerful way to stimulate the imagination of the viewers. For example, it is common to cut from 'couple kissing' to 'waves crashing on shore'.

Interestingly, while a great diversity of artists take advantage of the power of the sea on the imagination and emotional state of human beings, this power is used hardly in pornography. Whether you consider drawing, painting, sculpture, films or novels, when the genital organs are revealed, water is no longer employed as a symbol. The obvious reason is that true erotic art is not explicit, but suggestive. It is also notable that most of the well-known scientific researchers in human sexuality have not expressed a great interest in the effect of water on sexual arousal. Until now, the relationship between Homo and the sea does not appear as a serious topic among the current generation of mainstream scientists.

It is significant that the small groups of scientists who raised questions about the dozens of traits humans share with sea mammals, but not with other primates, have been marginalized.

Even if Eros thrives on mysteries, and even if human beings need mysteries, it seems unavoidable that in the near future we'll have to cope with the 'scientification' of the relationship between Homo and the sea. This new step will follow 'the scientification of love'[2] and the 'scientification of transcendence'[3]

One way to explore the future is to consider what we might learn from fast developing disciplines. The significant fluctuations of sea levels during the Palaeolithic Ages are already studied by small groups of highly specialized experts. When these fluctuations become common knowledge, we'll realize the narrow limits of what can be learned from fossil hunting. If most of our ancestors were living in areas that are now underwater, we'll probably never find their fossils. Studies of archaic humans—either Neanderthal or Sapiens—as skilled long-distance navigators cannot be dissociated from studies of the evolution of the oceans and of the climates. Furthermore, it is probable that, in the near future, population genetics will become the most authoritative discipline to understand how our ancestors have colonized the whole planet.

In a renewed scientific context, we'll learn to raise unusual questions.

What if my mother, in her poem, was expressing 'the original dream'? What if 'the original sin' is the reverse of the original dream?

2
Fluctuations of sea levels

As an interdisciplinary student in human nature, I have an innocent tendency to assume that the evolution of our planet and the evolution of Homo are two inseparable topics. In point of fact, up until now, it has not been so. In the countless books about human evolution, measurable criteria of the evolution of our planet, such as sea levels and average temperature, are rarely introduced.

It is useful to offer possible interpretations of situations that go against common sense. Although we might arbitrarily claim that studies of human evolutionary biology started with *The Descent of Man*, published by Darwin in 1871, in reality this discipline developed during the twentieth century. As for sedimentology, sedimentary palaeontology, and climatology, they also developed recently, in the age of overspecialization.

Is overspecialization making us blind? This is the implied question we are raising.

Measurable criteria

Our selected objective is to summarize measurable criteria of the evolution of our planet that all students of human nature should constantly keep in mind.

Interdisciplinary students in human nature must realize that during a period covering the last two million years (the Pleistocene) sea levels have fluctuated some 218 metres.

An estimated 20 million km² of territory was exposed on the world's continental shelves during the last glacial period (between 110,000 and 10,000 years before present). Exploitation of marine resources has undoubtedly created the potential for relatively high population densities and concentrations of archaeological sites. During this period some of the major phases of human history have taken place, including renewed human dispersals out of Africa.

Such facts suggest that until now palaeoanthropologists might have studied marginalized human populations that colonized inland territories, probably after migrations along rivers and around lakes.

Evaluations of changes of sea levels come from marine and estuarine documents about beaches and valley systems now submerged. For example, in Southern England, we have evidence of an amplitude of fluctuation relative to present level of more than 200 metres.[1]

In a renewed scientific and technological context, the time has come to raise questions about the future of 'underwater palaeoanthropology' as an emerging discipline. Preliminary findings may be presented, particularly from Western Australia.

When Homo arrived in Australia as early as 65,000 years ago, sea levels were around 80 metres lower than today.[2] After that, as the world plunged into the last Ice Age, which peaked around 20,000 years ago, sea levels dropped to 130 metres lower than today. Between 18,000 and 8,000 years ago the world warmed up. Melting ice sheets caused sea levels to rise. The sea level rise flooded 2.12 million square kilometres of land on the continental shelf surrounding Australia.

If we keep in mind that thousands of generations of human beings would have lived out their lives on these landscapes now under water, we realize the historical value of recently published studies of submerged archaeological sites off the Pilbara Coast in Western Australia. For several years, an interdisciplinary team of archaeologists, rock-art specialists, geomorphologists, geologists, specialist pilots and scientific divers on the 'Australian Research Deep History of Sea Country Project' have collaborated with the 'Murujuga Aboriginal Corporation' to find and record archaeological submerged sites.[3]

The characteristic archaeology of the region is dominated by open-air sites: especially engraved rock-art panels, but also includes: stone tool assemblages; quarries; circular or curvilinear stone structures interpreted as hut foundations or terraces to enhance trapping of sediment; standing stones of probable ceremonial significance; and shell middens.

One of the objectives of the published article was to describe the techniques, equipment and sampling procedures used in site discovery; it was also to analyze the results with reference to comparative data from sites on land.

Other investigations are under way in several parts of the world to explore the role of the coastal zones in population dispersal, to reconstruct these submerged landscapes, and to test their archaeological potential.[4]

We must mention in particular the case of Scandinavian and Baltic Sea regions.[5] In Norway, in the Hummervikholmen underwater site, it is notable that human skeletal remains are well preserved. It means that it is possible to find fossils of human beings who were living in places that are now submerged. Stable isotope analysis of the bones has revealed a heavy reliance on marine resources, and a DNA analysis has identified a genetic origin in populations from both Western Europe and Eastern Europe. In Sweden the situation is quite similar in that emphasis has been devoted to sites on land rather than to underwater investigations, apart from pioneer discoveries of underwater Stone Age finds in the Oresund Strait and the site of Havang. In Denmark, the site of Tybrind Vig offers a typical example of preservation conditions on underwater sites. It is one of the best known and best reported underwater excavations in Europe. A number of stone tools were found there. Eight excavations in Germany have produced abundant finds and details of preservation conditions, stratigraphic associations.

Among the other measurable criteria of the evolution of our planet, the first-place should be given to the huge fluctuations of the average global temperature. According to recent authoritative evaluations, the average temperature around 21,000 years ago was 9° Celsius. This is about 6° colder than the 2013-2017 average.[6]

Palaeomicrobiology

Palaeomicrobiology is the study of microorganisms associated with prehistoric material.

Since its introduction in the 1990s and its development during the past ten years after the 'microbiome revolution', this new discipline has made a big impact. Through the study of ancient DNA, using various tools and resources, researchers have gained valuable knowledge about the mechanisms of DNA preservation and gene evolution.

Let us recall that until recently bacteriologists could only look at microscopes and cultivate microbes on Petri dishes. They started to dramatically enlarge their horizon when they became geneticists, exploring the world of microorganisms by using the power of computer processing and new DNA sequencing technologies: not all bacteria seen under the microscope can be cultured, since their growth conditions are unknown. Bacteriologists can now see the 'unseen majority'. We have entered the era of 'metagenomics'. Our understanding of the entire living world has been revolutionized. The use of metagenomics has been fundamental in revealing the abundance and composition of marine microbial ecosystems.

Today, in a renewed scientific context, it is commonplace to present Homo as an ecosystem with a symbiotic interaction between the trillions of cells that are the products of our genes (the 'host') and the hundreds of trillions of microorganisms that colonize the body (the 'microbiome').

It is exciting to think about where the future of this field of study will take us when it becomes obvious that the evolution of our planet and the evolution of Homo are inseparable topics. One can assume that the microbiome of our Palaeolithic ancestors born in coastal areas was not programmed in the same way as the microbiome of inland hunter-gatherers. Can we already dream of a time when the studies of Palaeolithic humans include the field of marine microbiology?

The prototype of a modern interdisciplinary scholar

Allan Krill is originally a geologist, with a B.S. degree in Earth Science in 1976 from the University of California, Santa Cruz, and a Ph.D. in Geology and Geophysics in 1980 from Yale University. He is doing research on an impressive diversity of topics, such as number memorization techniques. He published documents explaining why Alfred Wegener's hypothesis of continental drift was rejected for so long.

I discovered Allan Krill as a unique interdisciplinary scholar because he has dared to publish an audacious hypothesis about the evolutionary relationship between humans and chimpanzees that takes into account his understanding of the relationship between humans and the sea.[7] His valuable hypothesis needs to be saved from oblivion because it is based on three kinds of established and essential facts:

— None of the non-human 260 species of living primates evolved large brains.
— There have been such fluctuations of the sea levels during the Palaeolithic eras that some territories have been occasionally islands, peninsulas or parts of mainland.
— Evolution of a new species by isolation on an island (peripatric speciation) is comparatively much faster than elsewhere.

Several hypothetical localities have been suggested as suited for such an evolutionary process. Leon La Lumiere considered the case of the Danakil island, in East Africa, in his study of the evolution of bipedalism.

Allan Krill focused on the Bioko island in the Atlantic Ocean. His hypothesis is valuable for two main reasons. First because Bioko is currently situated 32 kilometres offshore from South African areas populated by chimpanzees, and also because the basis of this hypothesis is that the enormous brain is the defining feature of Homo.

The work of Allan Krill is symbolic of the necessary revival of fertile interdisciplinary perspectives that have been stifled in the age of overspecialization.

<u>3</u>

Homo navigator

After suggesting that 'the original sin' might be the reverse of 'the original dream', and after recalling that during the Palaeolithic Ages a great number of human beings were living in areas that are now underwater, we are in a position to consider how our ancestors have colonized the whole planet.

We have at our disposal an accumulation of data confirming that archaic Homos were skilled long-distance navigators. We'll focus on the colonization of the Mediterranean Basin and the Pacific Rim.

We must remain cautious when trying to interpret available documents. The fluctuations of sea levels must constantly be kept in mind when there is proof of the presence of Palaeolithic Homos in territories that are currently islands. We must always first wonder if there have been phases in the history of sea levels when an island could be reached without travelling by boat.

The Mediterranean Basin

To justify and illustrate this caveat, we'll choose the case of Sardinia, which is, in the mind of modern people, a big island. The discovery of Palaeolithic workshops indicates a human presence in Sardinia in the period between 450,000 and 10,000 years ago.

According to researchers who found human remains at the 'Su Coloru cave' of Laerru in northern Sardinia (Anglona), some particularities of the lifestyle of these people suggest that they came by boat from the Italian coast.[1] However, we must be cautious before selecting one interpretation. There have been phases during the Palaeolithic era when it was possible to walk from the Italian coast in Tuscany to what is now Elba island. It was also probably possible to walk from Elba to the extreme North East coast of Corsica. Furthermore, there was no need for boats to go from Corsica to Sardinia and to cross what is now the Strait of Bonifacio, which is 11km long and never more than 100m deep. The oldest remains of the sapiens variety of Homo in Sardinia have been found in the central part of Sardinia in the 'Corbeddu cave' of Oliena.[2]

After choosing the case of Sardinia to illustrate the need to be cautious when considering plausible ways an island has originally been reached by humans, we'll use Crete as the opposite example, since it is an island that has undoubtedly been reached by boat. Crete lies approximately 160km south of the mainland and 40 kilometres from its closest neighbour. It has been an island for more than five million years. The first human settlement in Crete undoubtedly dates before 130,000 years ago, during the Palaeolithic Age.[3,4,5]

In excavations conducted near Crete's southwestern coast during 2008 and 2009, the team of Thomas Strasser unearthed hand axes at caves and rock shelters. Most of these sites were situated in an area called Preveli Gorge, where a river has gouged through many layers of rocky sediment.

At Preveli Gorge, Stone Age artefacts were excavated from four terraces along a rocky outcrop that overlooks the Mediterranean Sea. Tectonic activity has pushed older sediment above younger sediment, so 130,000-year-old artefacts emerged from the uppermost terrace. Other terraces received age estimates of 110,000 years, 80,000 years and 45,000 years.

These minimum age estimates relied on comparisons of artefact-bearing sediment to sediment from sea cores with known ages. Stone tools found there, archaeologists say, are at least 130,000 years old.

Today, archaeologists and experts on early nautical history are convinced that the surprising ancient navigators had crafts sturdier and more reliable than rafts. They also must have had the cognitive ability to conceive and carry out repeated water crossing over great distances in order to establish sustainable populations producing an abundance of stone artefacts. It is even not completely excluded that ancient navigators crossed the Mediterranean Sea and arrived directly from the North African coast, from what is now Libya.

Studies of early seafaring activity in the southern Ionian islands, off the west coast of Greece, have a double interest. They first confirm that Palaeolithic humans could reach by sea territories that have always been insular, whatever the fluctuations of sea levels. Furthermore, they focus on the Neanderthals as skilled seafarers.[6] The southern Ionian chain includes four large islands: Lefkada, Kefallinia, Ithaka and Zakynthos. On the map, the four islands appear as an arc-like string, separated from each other by straits of between 8.5 and 15 km.

In the southern Ionian islands, a total number of 15 Palaeolithic open-air sites have been detected. From the data that are now available, seafaring in the Ionian islands most likely started some time between 110,000 and 35,000 years before present. However, earlier beginning of seafaring activity as far back as 200,000 years before present are not excluded. It will probably remain difficult to have details about the boats used by Palaeolithic humans, since they were made of wood, so rotted away long ago. They probably had similarities with the oldest known Mediterranean boat, a dugout canoe from Lake Bracciano in Italy, which is 7,000 years old. It is notable that the artefacts found in the sites are characterized by a typical Palaeolithic-Mousterian industry. It confirms that Neanderthals had a seafaring culture for tens of thousands of years.

Questions from Gibraltar

The deep-rooted seafaring culture of Neanderthals must be taken into account to clarify uncertainties about the extreme western part of the Mediterranean Basin. Today, the Strait of Gibraltar is 14 kilometres long and its depth ranges between 300 and 900 metres. This means that even during the last major glaciation 20,000 years ago, when the level of the sea was lower by 110-120 metres, only navigators could travel from one continent to the other.

The Neanderthal exploitation of marine mammals on the Spanish side of the strait is well documented. The most significant documents were found in Vanguard and Gorham's Caves, that lie on the eastern side of Gibraltar.[7] These Palaeolithic humans, classified as Neanderthals, exploited seals and dolphins. They were undoubtedly navigators. If we add that the Moroccan coast was visible from the European side, it is highly probable that there were Neanderthals living on both sides of the strait. In fact, there is forgotten evidence that there were Neanderthals living on the other side of the strait, in Tangier. Excavations at Cape Ashakar (1939 to 1947) brought to light artefacts of Levallois-Mousterian type (particular to Neanderthals) and even a 'Neanderthaloid' tooth.[8]

Interestingly, it has been considered an established fact, since the 1960s and until the beginning of the twenty-first century, that Neanderthals had lived in Morocco. In 1960, miners had discovered a complete human skull in the wall of Jebel Irhoud cave, not far from the Atlantic coast. The skull was eventually handed over to the University of Rabat, who organized a joint French-Moroccan expedition to the site in 1961, headed by the French researcher Emile Ennouchi. This team of researchers found, in particular, parts of another skull and the lower mandible of a child. They did not hesitate to classify these hominins as Neanderthals after taking into account the elongated shape of the braincase and the signs of human habitation including an industry classified as Levallois-Mousterian, which is particular to Neanderthals.[9,10]

In 2004, the team of Jean-Jacques Hublin, a director at the Max Planck Institute for Evolutionary Anthropology in Leipzig, Germany, started to excavate the site again. They uncovered human bones relating to five individuals, including a jaw and skull fragments. They tried but failed to obtain DNA from the bones. They found that they had enough material to revise the conclusions of the previous researchers and expressed their conviction that the hominins who were living in Jebel Irhoud were Sapiens.[11] They offered vague interpretations of the elongated skulls. They were not interested in the typical Palaeolithic-Mousterian industry. They did not mention the excavations in Tangier in the early 1940s. Of course, at that time, they could not take into account that Neanderthals had been living on the European side of the Strait of Gibraltar and were skilled navigators. They were visualizing Morocco as a part of Africa, while it is also a part of the Mediterranean Basin nearly attached to Spain.

Hubin was given an opportunity to revise his disputable point of view in 2017 when a team led by archaeological scientist Daniel Richter and archaeologist Shannon McPherron, also at the Max Planck Institute, dated all the human remains found there to between 280,000 and 350,000 years old using two different methods. This was apparently in contradiction with the interpretations of Hubin, since the oldest Sapiens' fossils are 196,000 and 160,000-year-old skulls found in Ethiopia and DNA studies of present-day populations around the globe point to an African origin some 200,000 years ago. Instead of patiently reconsidering his interpretations, Hubin preferred to 'rewrite' our species' history. He conceived the theory of 'the pan-African origin of Homo sapiens'.[12,13] The basis of this theory is that early Sapiens dispersed around the continent and elements of human modernity appeared in different places and so different parts of Africa contributed to the emergence of what we call modern humans today.

I would not be surprised that this unconvincing theory is reconsidered in the near future. The ball is in the court of geneticists.

The Pacific Rim

It is worth noticing that the research teams who studied the colonization of the Mediterranean Basin by 'Homo navigator' are not the same as those who studied the colonization of the Pacific Rim. This implies that personal factors that might influence methodology and conclusions are eliminated and that we must give a great importance to the common points.

Before considering the history of the Pacific Rim during the past 100,000 years, we must open a parenthesis and keep in mind that hominid remains and stone tools dating back 700,000 to 800,000 years ago have been found on Flores island, Indonesia, by the joint Australian-Indonesian research team headed by Michael Morwood. According to Morwood, the mysterious Homo floresiensis got there by using boats.[14] We must also mention that the remains of another hominin have been found in the island of Luzon, 3000 km from Flores. The discoveries of 'Homo floresiensis' and 'Homo luzonensis' might intensify the hunt for more extinct hominins in this part of the world and help phrase new questions about the relationship between hominins and the sea.

The most undisputable evidence for the first use of boats in the South Pacific region is provided by the migration waves starting from Sundaland. Let us recall that one of the cradles of human civilization has been the prehistoric lowlands of the Southeast Asian peninsula. This area, commonly called the 'Sundaland', was above sea level during the last Ice Age. It was twice the size of India, and included what we now call Indo-China, Malaysia and Indonesia. There were migration waves towards the ancient continent of Sahul (which is now divided into Australian mainland, New Guinea and Tasmania), about 60,000 years ago.

It is now considered highly probable that the first peopling of Sahul by anatomically modern humans required crossings through Wallacea, which was made of a 970-km-long band of islands and ten ocean straits.[15] It seems likely that boats and coastal navigation existed for at least some time before that. It has been estimated that a founding population of between 1,300 and 1,550 of male and female individuals was necessary to maintain a low enough extinction threshold.[16]

It is of course difficult to know about the kind of food Homo navigator could consume in this part of the world. Fishing gear made from wood or plant fibres rotted away long ago. Even materials like shell, from which fish hooks were often made, have more often than not dissolved or crumbled to dust. However, remains of a variety of deep-water sea fish such as tuna (dating to 42,000 years before present) have been found in Jerimalai shelter in East Timor, as well as the earliest definite evidence for fishhook manufacture.[17] They are the proof of complex maritime technology. While some of our ancestors migrated from there towards the South, others migrated towards the North.

There is evidence that Palaeolithic Homo navigators were very active in the western zones of the Northern Hemisphere Pacific. It recently became a topical issue after explorers tried to unlock ancient mysteries about how our ancestors could be long-distance seafarers. A team of adventurers succeeded in paddling a primitive dugout canoe across more than 200 kilometres of ocean to demonstrate how Palaeolithic humans may have reached the Ryukyu Islands scattered between Taiwan and the Japanese mainland. They did not use modern navigational tools.[18]

As for the migrations towards the American continents, the sudden need to rewrite their history justifies, more than ever, the concept of 'Homo navigator'. The comparative importance of migrations along the coasts has probably been underestimated in the past.

Until recently, according to theories based on archaeological data, the ancestors of the indigenous cultures of the American continents had appeared in what is now New Mexico, where they developed the 'Clovis culture'. They were supposed to have reached the North American continent through an ice-free corridor that extended from Alaska to Montana. We have recently learned that life came to the ice-free Canadian corridor too late to sustain this theory.

It is undisputable now that there has been a more diverse set of founding populations of the continent than previously accepted.[19,20] Complex streams of gene-flow between North and South America have been previously unappreciated. Comments published in *Nature* summarize the current situation: 'Ancient genomics paints messy picture of America's first settlers.'

In such a context, the time has come to look at Homo as a skilled navigator and to accept that, probably, America's earliest humans did not arrive by land. It is worth keeping in mind the 'kelp highway' theory, as it was developed by Jon Erlandson and his team at the University of Oregon.[21] Let us recall that kelp is a type of large brown seaweed that grows in shallow, nutrient-rich seawater. There is a line of productive kelp forests that range along the coasts from northeast Asia to Baja California. Rich kelp-forest ecosystems could have supported seafaring people. There has not been abundant evidence for the 'kelp highway theory' because we would need the development of underwater archaeology to explore places that have been submerged by rising seas since the last glacial maximum.

The 'kelp highway' theory provides opportunities to forget the ice-free Canadian corridor theory and, at the same time, to reconsider mysteries, such as the reported findings at the Cerutti Mastodon site in Southern California.[22] According to the authors of the investigations, there is evidence that, 130,000 years ago, an unidentified variety of Homo was living there. These humans were apparently endowed with sufficient manual dexterity and experiential knowledge to use hammerstones and anvils. Critics have claimed that alternative interpretations have not been ruled out.[23] We must keep in mind, on the other hand, the point of view of the archaeologist Ruth Gruhn. Her examination of the broken bone fragments in the Cerutti Mastodon Site collection indicates that the hypothesis of breakage by modern heavy machinery is invalid, as a thick precipitate of soil carbonate on the broken surfaces proves that the breakage was very ancient.[24]

While the 'kelp highway' theory can offer plausible interpretations for the colonization of North and Central Americas, it remains difficult to interpret highly mysterious data regarding South America. Genetic studies have indicated that some Amazonian Native Americans descend partly from a Native American founding population that carried ancestry more closely related to indigenous Australians, New Guineans and Andaman Islanders than to any present-day Eurasians or Native Americans.[25] This signature is not present to the same extent, or at all, in present-day Northern and Central Americans.

The most intriguing mystery on which we should focus is the case of Monte Verde, an archaeological site in Southern Chile. Monte Verde provides an opportunity to radically reshape the way we think about the earliest inhabitants of the Americas.[26] Radiocarbon dating has provided a date of 14,800 and possibly 33,000 before present, establishing Monte Verde as the oldest-known site of human habitation in the Americas (as long as the findings at the Cerutti Mastodon are not confirmed). The main issue is not the distance between Alaska and Southern Chile, which is in the region of 13,000 kilometres. We must first think in terms of latitude: how long would it take for a population originally adapted to the climate of Alaska (about 40° North) to migrate to South Chile (about 39° South)? Such human groups would have to adapt to temperate climates of North America, before adapting to Northern tropical, equatorial, Southern tropical, Southern temperate latitudes and finally the climate of the extreme South of the American continents. It is radically different from the migration of our ancestors from North East Africa to South East Asia, with distances evaluated mostly in terms of longitude.

At such a turning point in our understanding of the colonization of the planet, we'll dare to suggest that the wacky and even apparently absurd hypothesis of a human migration towards the coasts of South America via the southern part of the Pacific Ocean should not be radically dismissed. We must keep in mind that 20,000 years ago, when the sea levels were more than 100 metres lower than today, there were countless islands in the Southern Pacific Ocean. Some of them still exist: Pitcairn Islands, Easter Island, Sala y Gomez, Desventurados Islands, San Felex, San Ambroso, Alejandro Selkirk and Robinson Crusoe, in particular.

We must also keep in mind what is known about how early navigators were guided in unchartered seas and the important part played by birds.[27] It is significant that the Rapa Nui name for Sala y Gomez island means 'Bird's islet on the way to Hiva'. The strong spatial skills of birds such as the terns remain mysterious. Even in historical times, explorers were still guided by birds. Vincento Pinzon, the Spanish navigator who sailed with Christopher Columbus on their first voyage to the New World, was quoted as saying: 'Those birds know their business.' Is it plausible that archaic navigators had reach the highly productive ecosystem related to the Humboldt Current, which runs along the South American coast? Still around the Pacific area, we must keep in mind that the Marshall Islands, that lie more than 2,000 miles from the nearest continent, were discovered and colonized about 3,000 years ago by seafarers from South East Asia who were relying on the art of 'wave piloting'. Wave pilots steer by the feeling of the ocean itself: before the age of stick charts, they could perceive disruptions in ocean swells by islands. It is probable that ancestral humans have also used a combination of wind and stars to navigate. We may be amazed by the capacity of Homo navigator to plan extremely daring voyages.

We have the proof that the canoe squadrons of archaic Homo were carrying men and women since, after reaching new territories, they could reproduce and start off populations. This fact can inspire questions about gender differences at a time when the concept of male superiority in spatial thinking is challenged.[28,29] It seems more appropriate today to think in terms of advantageous complementarity between male and female brain functions.

From Madagascar to Easter Island

These audacious behaviours can be explained by a conception of the world according to which the ocean is full of islands, and the sea is the opposite of an obstacle: it is the link between territories. This is the best key to interpret linguistic mysteries. How to explain that languages of the 'Austronesian' families are spoken in places apparently as diverse as, for example, Madagascar, Borneo, Taiwan and Easter Island? Look at a globe: the sea is the obvious link.

Presenting Homo navigator as one of the prototypes of prehistoric Homo should renew our curiosity about human nature and the limits of human adaptability.

<u>4</u>

Looking Towards The Future?

As interdisciplinary students in human nature, we are combining records of the past in order to look towards the future. While there was only one past, it is revealed to us in many ways. None of the different records represents an unbiased picture of the whole.

Limited perspectives

Each perspective is in the hands of specialized researchers and has limited objectives. Archaeologists look at physical objects that have been shaped by human contact. These include not only tools, ornaments and pottery, but also soils and waste deposits. Tools produced by early human ancestors are difficult to distinguish from naturally occurring objects. For obvious reasons, in archaeological records, many stone tools may be found, but not wooden ones. Furthermore, until now, most archaeologists do not take into account that during the Palaeolithic phases of human history, a great number of human beings were living in places that are now underwater.

During the twentieth century, palaeoanthropologists—who focus on the fossils of humans—played a central role. Although it is commonplace today to associate the work of fossil hunters with the African continent, it is worth recalling that Asia was the cradle of palaeoanthropology. China was considered a promising home for early humankind. After the

German palaeontologist Max Schlosser described in 1903 a human tooth from Beijing, there were reasons to believe that future work would discover a new human being in China.

The turning point took place when Raymond Dart, in 1924, discovered, in South Africa the first 'Australopithecus Africanus'. A well-preserved juvenile skull had been found in a limestone quarry at Taung. Dart recognized an early human because the brain dimensions were too large for a baboon or a chimpanzee. As Dart was not part of the scientific establishment, and because he found the fossil in Africa, and not Europe or Asia, where the establishment supposed man's origins, his findings were originally dismissed. Another 20 years passed before Dart's claims were taken seriously. After that, the real hub of palaeoanthropological studies was in East Africa.

We must constantly keep in mind the narrow limits of what we can learn from palaeoanthropologists. When human fossils are found it means first that the geological and environmental conditions are absolutely exceptional. Many geological conditions have to occur together so that a hominid fossil has even a tiny chance of being detected one day by a scientist. For example, in the case of the Olduvai Gorge in Tanzania, one of the most important palaeoanthropological sites in the world, periodic flows of volcanic ash from Olmoti and Kerimasi helped to ensure preservation of the fossils. The remote Denisova Cave in the Altai Mountains in Southern Siberia is another typical example of a paradise for palaeoanthropologists, because the average temperature inside the cave is 0° C. It is as if precious fossils had been kept in a freezer. It is not by chance that remains of Denisovans were found there and that the first high coverage genome of Neanderthals was taken from a toe bone found in the same cave.[1] It does not mean that a comparatively great number of Denisovans or Neanderthals were living in Southern Siberia.

When considering the limits of what we can learn from palaeoanthropologists, we must also recall that only hard parts of the body such as the teeth, skull, jaw and long limb bones can be preserved and found in fossils. In addition, fossils are often distorted by earth movements during their long burial. Because many adaptative characteristics concern the soft tissues only, it is impossible, for example, to distinguish the fossils of a tiger from the fossils of a lion.

To these well-known limits of what we can learn from fossil hunters, we must add, once more, the deep-rooted tendency to ignore the significant fluctuations of sea levels during the Palaeolithic Ages, during the millennia when our ancestors have colonized the whole planet. Meanwhile, as long as underwater palaeoanthropology does not develop, it is commonplace among highly specialized researchers to think and to behave as if most human beings were living inside the continents before the Neolithic revolution.

Although genuine palaeoanthropology has limits, its field has enlarged during the second half of the twentieth century thanks to the strong links with emergent or fast-developing technics and disciplines.

This is the case of fossil-dating methods, such as radiocarbon dating. Living plants and animals absorb carbon from the atmosphere, including carbon-14—a radioactive form of the element produced when cosmic rays from the sun interact with nitrogen in the upper atmosphere. But when organisms die, they no longer take in any carbon, and the carbon-14 in their bodies begins to decay at a known rate. Scientists measure the amount of carbon-14 in biological materials to determine when that organism died. The method was developed in the late 1940s at the University of Chicago by Willard Libby. Because the half-life of carbon-14 (the period of time after which half of a given sample will have decayed) is about 5,730 years, the oldest dates that can be reliably measured by this process are approximately 60,000 years ago. Potassium-Argon and Argon-Argon, on the other hand, can date fossils between 250,000 years ago and 4 billion years ago. The gap is precisely when modern humans were evolving... often referred to as 'the muddle in the middle' because of difficulty dating many sites.

A unique dominant status

While fossil-dating methods offer examples of technics at the service of palaeoanthropology, human evolutionary genetics has already reached a unique dominant status. It is suddenly experiencing such massive advances that from now on it provides keys to address fundamental questions about the past and the future of Homo. In the age of spectacular advances in genetics, and although DNA cannot always be extracted from fossils, a renewed understanding of evolutionary history shapes our expectations about the future.[2]

It is impossible to present a catalogue of all the pending questions about our species that might be clarified in the near future thanks to evolutionary genetics. In the framework of our interest in particular aspects of human nature, it is essential to wonder how deep rooted is the need to explore the seas. We have new reasons and new ways to phrase unusual questions at a time when modern Homo may be presented as the result of genetic admixture between different variants of our species. Ancient DNA analysis has been successful in generating a draft of Neanderthal genome sequence and a high coverage Denisovan sequence. Analyses suggested that their genetic lineages diverged from those of modern humans about 800,000 years ago.

We are reaching a time when we can unhesitatingly claim that Neanderthals had a seafaring culture. It is obvious when considering, in particular, how Homo has colonized the Mediterranean Basin. It is also undeniable that Denisovans have been able to cross large bodies of water to migrate to territories beyond the Wallace Line such as Wallacea (a group of islands separated by deep-water straits from the Asian and Australian continental shelves) and also Sahul.[3] Let us recall that the Wallace Line is a faunal boundary line, drawn in 1859 by the British naturalist Alfred Russel Wallace, that separates the biogeographical realms of Asia and a transitional zone between Asia and Australia. West of the line are found organisms related to Asiatic species; to the east, a mixture of species of Asian and Australian origin is present.

If we keep in mind that remains and stone tools dating back 700,000 to 800,000 years ago have been found on Flores island, Indonesia, and that the mysterious Homo floresiensis probably got there by using boats, and if we also mention the remains of another hominin found in the island of Luzon, 3,000 km from Flores, we can conclude that exploring the seas is a deep-rooted need among humans.

At a time when the prehistorical lowlands of the Southeast Asian peninsula appear as a cradle of human civilization, and at a time when there is a sudden need to rewrite the history of the migrations towards the American continents, it is imperative to train ourselves to think like population geneticists and to become familiar with their vocabulary. To follow probable imminent fast advances in the genetics of population, we all need to become familiar with concepts such as introgression (the incorporation of genetic material from one species or subspecies into another one) or adaptative radiation (when organisms diversify rapidly from an ancestral species into a multitude of new forms).

Why the Americas before Madagascar?

After presenting an accumulation of data confirming that both Neanderthals and Denisovans were skilled long-distance navigators, it is paradoxically more difficult to consider the case of the 'pure' Sapiens, who migrated 'recently' from Africa to Eurasia. We'll offer reasons to wonder if Sapiens became more curious about 'distant islands' after incorporating genetic material from Denisovans or/and Neanderthals. We must keep in mind that genomes of humans in the southwestern part of the Pacific Rim contain less variations than those in Africa, but have incorporated genetic material from archaic Homos.[4]

To suggest possible answers to this question we'll focus on plausible effects of Sapiens-Denisovan interbreeding. We have at our disposal a huge valuable study about modern humans who have inherited genetic material from Denisovans.[5] This study assembled data from 33 populations (using the 'Single Nucleotide Polymorphism Database'). It appeared that the incorporation of genetic material from Denisovan occurred within the southwestern part of the Pacific Rim rather than on the Asian mainland.

The results of such a study provide keys to interpret mysteries, such as the mysterious case of Madagascar. While Madagascar is very close to the East coast of the African continent, it is nearly established now that the first settlers arrived from Borneo in coast-hugging craft that skirted the Indian Ocean. Genetic studies from southeastern Borneo populations revealed an origin of the ancestry of Malagasy among one particular ethnic group (the Banjar).[6] It is highly significant that some 90% of Malagasy vocabulary is from the language of the Ma'anyan, an indigenous group of roughly 70,000 people who live in remote inland areas of southeastern Borneo. As for the Bantu, from South Africa, they did not move to Madagascar until 1,100-700 years before present.[7]

The point is that, until now, the appropriate question has not been phrased properly: how to explain that the originally African Sapiens settled in places such as Australia, Japan and the Americas long before reaching Madagascar? If phrased properly, this kind of question is easily enlarged. We might wonder, for example, if the relationship of the pure African Sapiens with water is modified by the incorporation of genetic material transmitted by archaic Homos. How can a genuine Sapiens acquire the need to explore the seas or become a long-distance navigator?

We might raise similar questions about the islands in the Atlantic Ocean that are close to the African continent. It is notable that the Cape Verde archipelago appears not to have been inhabited by humans until the fifteenth century, when Portuguese explorers discovered and colonized the islands. As for the Canaries, it seems established that they were first inhabited by the Guanches of Tenerife, Berber-related people with a Neolithic culture.

Underwater fossil hunting

Since the glacial maximum, a landmass the size of South America has been inundated worldwide. On the day when this piece of information becomes common knowledge, underwater fossil hunting (i.e., underwater palaeoanthropology) will probably become an irreplaceable complementary key to studying human nature.

Until now only a very small number of underwater human fossils have been found. Interestingly, it was never in the framework of organized scientific research.

This is the case of the skull of a Neanderthal, dated 60,000 years ago, fortuitously found in 2001 by Luc Anthonis at the bottom of the North Sea, about 16km from the Dutch coast. Luc Anthonis was a private Belgian collector.[8]

It is worth recalling that, in 1994, a cabin owner at Hummervikholmen, southernmost Norway, found human skeletal remains in his small harbour while removing sediments. Human skulls and limb bones were flushed out of the seafloor when the owner was deepening his small harbour with the aid of the water stream of the propeller of his outboard motor. The find is comprised of the very well-preserved bones of at least three, but perhaps as many as five, individuals. One skull is from a female, about 35–40 years of age at death. It is robust, with a high upper face and rather low eye sockets, resembling other Scandinavian Mesolithic female skulls. A second skull is from a man of approximately the same age. Analyses (δ^{13}C and δ^{15}N measurements) indicate that more than 80% of the dietary protein of the individuals was derived from marine mammal resources.[9]

There are reasons to explore the Atlantic European coastal areas, particularly in Wales. Humans have lived in Wales for over a quarter of a million years. Neanderthals preceded modern humans. Evidence for the first human occupation of Wales can be found in caves close to the coast. These caves are found in limestone areas, for example in parts of Denbighshire, Gower and south Pembrokeshire.[10]

Whatever the technological advances, there will always be immutable limits to the likelihood of finding fossils of mammals in tropical climates and in erosional coastal areas without sediment deposition. Some of these limits are difficult to evaluate. For example, there is a lack of information about the comparative role of 'bone-eating worms' in the destruction of skeletons of marine vertebrates at different latitudes.[11]

These considerations about the future of underwater palaeo-anthropology are first and foremost an opportunity to suggest that the concept of interdisciplinary perspective will probably enlarge in the near future.

5

Homo's Best Friends

'Tell me who your friends are and I'll tell you who you are.' This is the usual translation of the French aphorism *'Dis-moi qui tu hantes et je te dirai qui tu es'*.

This is why our study of human nature must include considerations about dolphins, the best friends of Homo in the seas, and dogs, the best friends on land.

The dolphin connection

The human-dolphin connection is as old as our historical memory and is cross-cultural.[1] How did it come about?

Traditional fishing cooperation might be at the root of this connection. Until recently in places such as Mauritania, for example, there was actual complicity between fishermen and dolphins, with both sides helping to round up the catch. It seemed as if they had struck a deal to share the fish. In view of this close relationship, it is not surprising that in that culture killing a dolphin was tantamount to killing a man. Similar fishing complicity has been reported in ancient times in the Mediterranean by such people as the Greek author Oppian and the historian Pliny. Coastal-dwelling Aborigines in different parts of Australia have also fished in harmony with dolphins since time immemorial. This may help to explain the 'Monkey Mia' phenomenon. Monkey Mia, on the west coast of Australia, is the only place in the world where children can easily meet families of friendly dolphins in shallow water.

The human-dolphin connection seems to be based on mutual trust. Using their powerful snout like a ram, dolphins could easily kill a man, but they don't do it. On the other hand, they wouldn't hesitate to kill a shark threatening a baby dolphin.

Dolphins never attack humans, but there have been countless stories of them saving lives. In 1993, for example, two men and a boy, set out in a small boat from Le Conquet in order to gather shells on a nearby islet. Too late they noticed that the weather was changing. Heading northeast, they tried to get home but the sea was so rough that the small craft was in imminent danger of capsizing. Then a school of dolphins appeared and clung to the craft as if trying to make it stable. After a while, and to the sailors' bewilderment, the dolphins started pushing the boat south, towards Camaret. Eventually all became clear when they realized that the dolphins had led them into calmer waters, probably saving their lives.[2]

Such a story might be regarded as pure fabrication had there not been tales of dolphins saving men from drowning and steering ships to safety reported by people all over the world and through the ages. Documented examples appear in such diverse places—classical Greek texts and Polynesian tales, for example—that no one could claim collaboration or a conspiracy to mislead.

Whatever aspect of the human-dolphin connection we consider, the cross-cultural similarities are striking. In places as far apart as possible on this planet there are legends of humans changing into dolphins and vice versa. According to Greek mythology, dolphins were originally humans who lived in cities along with mortals before exchanging the land for the sea and adopting the form of fishes. The beliefs of the Maoris in New Zealand are basically the same, and their tradition has it that a dolphin is a 'human being of the sea'. Similarly, the natives of an island in the Gulf of Carpentaria, northern Australia, believe that humans and dolphins have a common ancestor: she is the mother Ganadja, who one day took the shape of a human being. The natives of another island in the same gulf call themselves the 'Dolphin People' and believe that their shaman is a dolphin who has chosen to reincarnate as a human being. In fact, the shaman-dolphin connection is another striking cross-cultural phenomenon. More generally speaking, dolphins have always been seen as divine creatures or divine guides for humans. Given the separateness of their origins, all these beliefs become highly significant when they are brought together.

The ancient and apparently universal interest in dolphins means that the current fascination for them cannot be seen as transitory or merely fashionable. Of course, the recent resurgence of interest does owe something to publicity about modern fishing practices which endanger the species and provoke outrage around the world. But it also suggests that humanity is suffering from a guilty conscience at transgressing deep-rooted moral rules.

The most spectacular modern form of fascination for dolphins is the desire to swim with them. As Amanda Cochrane and Karena Callen pointed out in their book *Dolphins and Their Power to Heal*: 'the extraordinary experience of swimming with wild dolphins fired a fascination which turned into a near obsession for us both'.[3] Others whose relationship with dolphins have inspired widespread interest include the French diver Jacques Mayol, subject of the film *The Big Blue*, and internationally known 'dolphin people' such as Estelle Myers and John Hunt. The desire to swim and dive among dolphins has been enhanced by factors as diverse as the writings of experts such as John Lilly[4] and Horace Dobbs[5,6], by television programmes, by disapproval of dolphinariums and, not least, by opportunistic travel agents.

Meeting dolphins in their natural environment is a new form of tourism which is developing dramatically in places such as Kaikoura in New Zealand. Increasing numbers of humans and dolphins are involved in these encounters, organized by companies which need a permit from the government. The strategy of these companies is based on the works of dolphin expert Wade Doak. The main step is to establish a trusting relationship between a crew and a group of dolphins through regular meetings, during which tacit rules are established and interest is stimulated by creative games. In Kaikoura the dusky dolphins were the first involved. Now bottlenose dolphins and Hector's dolphins join the parties. Encounters with wild dolphins have also been organized in the Bahamas and in the Gulf of Aqaba in Israel.

Encounters with dolphins need not be restricted to the framework of tourism and holidays. They can also be encountered in a healing context. In 1972 Dr Henry Truby, from the World Dolphin Foundation in Key Biscayne, Florida, observed that neurologically impaired children had exceptional responses to close contact with free-swimming dolphins. At the same time Betsy Smith, a specialist in autism at the University of Florida, noted unexpected positive responses from the interaction between a usually aggressive, unruly adolescent dolphin and a retarded adult called David.[7] While the dolphin became gentle, patient and attentive, David, who was usually very cautious near the water and slow to adapt to new stimuli, immediately waded in and began talking with the dolphin, stroking him and playing with him.

Dr Truby, Betsy Smith and another expert in autism met and laid the foundation of a research project. The place was to be a deep artificial cove in a canal in Key Largo, from which the dolphins could return to the ocean when they wanted. Here, autistic teenagers could be in contact with the dolphins, either sitting on a platform where the dolphins would be playful with them, or entering the water itself. The first observation had been that the dolphins persevered in trying to communicate with autistic children, remaining friendly and joyful even when faced with unresponsiveness, which ordinary pets wouldn't do. At several conferences on the human-animal bond Betsy Smith presented positive conclusions. One of the most significant was that the encounters and their after-effects were found to be life-enriching by the parents of all who participated.

It is also in the 1970s that dolphin enthusiast Horace Dobbs in Great Britain became intrigued by the sudden recovery—after swimming with a dolphin—of a lifeboat mechanic who was close to a nervous breakdown. A few years later in Wales, while leading a group of people who wanted to meet a dolphin, Dobbs noticed that the dolphin spent most time with a man who was in a state of deep depression and had heart disease.

After several similar observations, Horace Dobbs set up a series of projects, including Operation Sunflower, the aim of which was 'to study friendly and ethical ways to take advantage of the healing power of dolphins, without exploiting them and disturbing their usual way of living'.

Another programme, designed by Dr David Nathanson in the USA, is based on using dolphins and music to capture the attention of children with intellectual disability. Such a programme, however, implies that the dolphins would have to be trained. The future of dolphin therapy depends on how ethical questions are posed and answered.

Several theories have been put forward to interpret the healing power of dolphins. For example, it has been assumed that dolphins can sense imbalances in subtle bioenergy fields and can influence them. It has also been suggested that swimming with dolphins may trigger alpha brain waves, which are linked with serenity. Other theories stress the healing potential of sounds emitted by the dolphins. This was reflected in the work of swimming instructor Denis Brousse in Lyons, who was using sounds similar to those chanted by Tibetan monks when he was in the water with neurologically disordered children.

In fact, whatever the method of interaction, the healing effect of dolphins is not wholly surprising; it might even be anticipated if what is already known is taken into account. For example, we know that to take an initiative is health enhancing, while a state of submission is the typical disease-creating situation. For this reason, a therapy chosen on the initiative of the patient has a higher probability of working.

This attitude probably also accounts for the success of alternative therapies at the very time when pharmacology and surgery are more effective than ever. Let us recall the experiments with rats receiving electric shocks: what made the rats ill is not the number or intensity of the shocks but the fact that, when receiving such shocks, they could neither fight with a companion nor escape—they had to submit. We also know about the therapeutic power of water and the deep-rooted cross-cultural fascination for dolphins compared with other sea mammals. We know too that dolphins are joyful animals and that joy is contagious. Moreover, there have been countless studies demonstrating the positive influence of animals on our health, especially when the human being and the animal are known to each other. This is a point which has been stressed by several therapists: 'dolphin therapy' seems to be more successful when the dolphin and the human have established a relationship.

The friendly and inquisitive nature of dolphins makes us warm to them and give them human names, just like real friends. Some of the most famous dolphins were Jean-Louis in Brittany, Jojo in the Bahamas, Percy in Cornwall, Fungie in Ireland, Simo in Wales, Freddie near Newcastle, Jotsa in Montenegro and Dolphy in the South of France. Personalizing our relationships with dolphins is no different from humanizing them.

The canine connection

Dogs were already friendly companions of human beings in Palaeolithic Ages, long before the domestication of animals became one of the facets of the domination of nature. According to the evolutionary geneticist Elinor Karlsson, 'dogs probably started exploiting humans because they were a useful resource that helped them to survive'.[8] In other words, dogs might have moved freely, following humans or moving between groups when it suited their interests.

A recent huge study of ancient dog DNA traces canine diversity to the Ice Age. About 11,000 years ago, before any other animal had been domesticated, there were already five different types of dogs with distinct genetic ancestries.[9] In the current scientific context, we can therefore present dogs as old and close animal partners, although we cannot say when and where this deep relationship began. By testing 375 eight-week-old dog puppies, the team from the Arizona Canine Cognitive Center has demonstrated that dogs are genetically prepared for communication with humans.[10] The time has come to realize that studying these animal companions adds another layer to our understanding of the history of our species: it is probable that our ancestors used to bring their dogs with them as they migrated across the world.[11] In the age of genetics, the canine connection is becoming a complementary and fruitful way to reconsider the colonization of our planet. Advances in the isolation and sequencing of ancient DNA can reveal the population histories of both people and dogs. It has already been possible to compare population genetic results of humans and dogs from Siberia, Beringia, and North America. There is a close correlation in the movement and divergences of their respective lineages. It appears, from several studies, that dogs accompanied the first people into the Americas and travelled with them.[12,13]

<u>6</u>

The Super Brainy Mammals: Homo and Tucuxi

In ancient Greece, Aristotle had already underlined that 'of all the animals, man has the largest brain in proportion to his size'. Today the brain mass in relation to the size of the body is usually quantified in terms of 'encephalization quotient' or 'brain-to-body-weight ratio'. Among humans, the encephalization quotient (about 7.4 to 7.8) is mysteriously three times higher than among the other members of the chimpanzee family (about 2.2 to 2.5).[1]

Nutritional needs

This significant discrepancy must be reinterpreted at a time when the specific nutritional needs of the human brain, and particularly the developing brain, have been clarified. It is understood today that a molecule of fatty acid commonly called DHA is essential to feed the brain as a fatty organ. DHA is a very long-chain (22 carbons) highly unsaturated (6 double bonds) molecule that belongs to the omega 3 family. Let us just mention that 50 per cent of the molecules of fatty acids that are incorporated into the developing brain are represented by DHA. The important point is that this molecule is preformed and abundant in the seafood chain only. Human beings who don't have access to the seafood chain must rely on the parent molecule of the omega 3 family (alpha linolenic acid: only 18 carbons and 3 double bonds) and transform it thanks to their enzymatic system and the help of catalysts (mostly minerals).

Among the other 'brain specific nutrients', we must give the first place to iodine because of its essential role in thyroid hormone production, which, in turn, is needed for normal brain development and brain functions. Iodine is scarce on the earth's surface because over hundreds of millions of years it has been washed away by rain and glaciation and transported from the terrestrial crust to the sea. Seawater contains 10- to 200-fold higher quantities of iodine than freshwater.[1] In the sea, algal phytoplankton, the basis of the marine food chain, acts as a biological accumulator of iodine.

It is notable that iodine is the only nutrient for which governments legislate supplementation, so that iodination of table salt is mandatory in many countries (the iodine of sea salt disappears during the process of desiccation in salt marshes). In spite of such widespread legislation and many public health strategies (such as dripping iodine in the water of Chinese irrigation ditches), iodine deficiency remains a common nutritional deficiency at a planetary scale. It is the leading cause of preventable intellectual disabilities.

We must seriously take into account the increased needs during pregnancy and lactation. According to a British study, a daily supplement of iodine when the mother was pregnant is associated with a 1.22 average increased IQ in childhood.[2]

These considerations about the encephalization quotient of primates and the specific nutritional needs of the brain suggest that the main differences between Homo and all the other primates probably took place during a phase of adaptation to coastal areas and easy access to the seafood chain. There are serious reasons to keep in mind the audacious hypothesis by Allan Krill, that is based on three essential factors: the fluctuations of sea levels during the Palaeolithic eras, the nutritional needs of the brain, and the fact that isolation on an island tends to speed up the process of evolution. If our ancestors reached a very high encephalization quotient during a phase of easy access to seafood, we must wonder how, during ensuing phases of evolution, they have adapted to a great diversity of diets, including diets without seafood. To be in a position to phrase appropriate questions about humans, we'll consider in parallel the particular case of mammals of the dolphin family.

The dolphin family

Homo is not the only 'super brainy mammal'. Members of the dolphin family also have an encephalization quotient above 4. Furthermore, some of them, particularly the tucuxi, left the marine environment some million years ago and adapted to life in rivers. There are therefore inescapable reasons to conjointly study the chimpanzee family and the dolphin family.

It is worth emphasizing that the tucuxi is probably second to modern human in terms of encephalization quotient (4.56 for tucuxi versus 4.14 for bottlenose dolphins).[3,4] It lives only in fresh water in the wide reaches of the Amazon River. In general, scientists are confused about how different types of dolphin are related to one another, even, for example, bottlenose and spinner dolphins. However, in the particular case of tucuxi, it is established that it has the same ancestor as the Guiana dolphin, also known as the estuary dolphin, which is found in the waters of the eastern coasts of South and Central America.[5] The tucuxi probably evolved around 2.3 million years ago, heading inland away from the common ancestor, when the Amazon River first flowed around, due to a combination of the uplift of the Andes and a worldwide drop in sea levels as the first Ice Age loomed. The tucuxi is probably as old as the river it swims in.

It might be fruitful to combine two apparently unrelated questions. The first one is: what enabled dolphins to adapt to life in rivers? The second one is: how could the big brain Homo adapt to life far from the sea?

Starting from these questions we must first mention the need for comparative studies of the unsaturated fatty acid biosynthetic process, and particularly studies of desaturase enzymes that regulate unsaturation of fatty acids through the introduction of double bonds between defined carbons of the fatty acid chains.

Too rational to survive

In the future, to try to understand the fluctuations of brain size in our species, we might start from unusual perspectives. The point of departure might be the case of humans with larger brains than modern Homo. This might help us to change our way of thinking and challenge the simplistic vision of an 'upward trajectory', according to which human brains became bigger and bigger throughout the evolutionary process.

It seems that most of us prefer to ignore that some previous humans had been bigger-brained than modern-day humans and that, above a certain threshold, an excessive brain development might be a handicap in terms of survival of our species.

We must first recall that a modern human brain averages 1,350 cubic centimetres. The brain of Neanderthals was around

1,500 cubic centimetres. Neanderthals have disappeared. Modern Homo has colonized the whole planet. It is important to keep in mind that fossils of hominids with brains above 1,600 cubic centimetres have been found.[6] This is the case of 'Amud 1', a nearly complete male adult Southwest Asian Neanderthal skeleton thought to be about 55,000 years old. It was discovered at Amud in Israel. Its cranial capacity was around 1,740 cubic centimetres.[7] The skeleton of the famous 'Boskops Homo', found in South Africa by two farmers, is also characterized by a cranial capacity above 1,700 cubic centimetres.[8] Fossil skulls found at Wadjak, Indonesia, and at Fish Hoek, South Africa, also have 1,600 cubic centimetres capacities. Dozens of skulls from Europe, Asia, and Africa exhibit similar huge size, including skulls that were found in the caves of Cro-Magnon, in southwestern France.

We are at a time when humans number in the billions. We are also at a time when the upper limit in brain size has dramatically decreased. What are the possible links between these two facts? One possible interpretation is that some humans developed huge brains during the Palaeolithic Ages when the sea levels were significantly lower. They became adapted to precise locales, like typical mammals: lions are adapted to grasslands and savannas, polar bears live on ice, etc. Groups of humans adapted to coastal areas. The specific nutritional needs of their gigantic brains made them highly dependent on the seafood chain. They had reached a state of reduced adaptability in terms of nutritional requirements. This is one of the probable reasons why a gigantic brain may be a handicap.

Another probable reason is a correlation between a huge brain and an excessive degree of rationality. The will to survive and the struggle for life are not based on reason or logic.

The many facets of love are outside the field of rationality. We survive as individuals, as groups or as a species because we are not purely rational beings.

Rationality is related to the activity of the part of the brain—the new brain or neocortex—that is gigantically developed among humans only. The neocortex appeared originally as a tool at the service of vital physiological functions. Homo is a mammal whose neocortex is strong enough to exceed its role as a tool, and the concept of neocortical inhibition is becoming a key to understanding human nature.

We must give their real importance to data supporting evolutionary theories of human suicide, according to which a threshold intelligence is necessary for self-damaging behaviour (expression of 'impaired capacity to love oneself'). According to a study comparing 85 countries, suicide rates are related to the average intellectual development.[9] Furthermore, excess suicide prevalence has been observed in the highly gifted.

If, in the future, underwater palaeoanthropology can develop in some parts of the world, it might become possible to evaluate what the proportion of humans with enormous brains was in populations adapted to easy and quite exclusive access to the seafood chain.

Meanwhile, it is difficult to explain how we became the dominant creature on planet Earth.

7

From Aesop to Elaine Morgan: What pioneers have in common

Those who know about the specific nutritional needs of the brain, those who are aware of the spectacular fluctuations of sea levels during the Palaeolithic Ages, and those who are interested in the way our ancestors colonized great parts of the planet, are surprised that during thousands of years countless scholars pronounced on human nature without focusing on the relationship between Homo and the sea. This is a reason to mention a comparatively small number of pioneers.

Aesop

We'll start with *Aesop*, as the author of the fable 'the Monkey and the Dolphin'. A ship was wrecked off the coast, close to the port of Athens. The dolphins took the wrecked people on their backs and swam with them to shore. When one of the dolphins saw a monkey struggling in the water, he thought it was a man and made the monkey climb up on his back. Then off he swam with him towards the shore. On the way the dolphin turned his head and realized his mistake. Without more ado, he dived, left the monkey to take care of himself and swam off in search of some human beings to save. Through this fable, Aesop wanted to emphasize that it is electively between humans and dolphins—rather than between primates and dolphins—that there are powerful compassionate links.

Of course, when we mention Aesop, we refer to the legendary author of innumerable fables mixing an impressive diversity of topics. We do not try to know who, in reality, was at the root of each fable. Referring to Aesop and the great diversity of fables associated with his name is first an opportunity to focus on the keyword 'interdisciplinarity'.

Max Westenhöfer

From ancient Greece, we'll jump directly to the middle of the twentieth century.

In 1942, during the Second World War, the German pathologist *Max Westenhöfer* found a way to publish a book about various human characteristics (hairlessness, subcutaneous fat, the regression of the olfactory organ, webbed fingers, flat-footedness, direction of the body hair, etc.) that could have derived from an aquatic past. He stated that the postulation of an aquatic mode of life during an early stage of human evolution is a tenable hypothesis, for which further inquiry may produce additional supporting evidence.[1] For obvious historical reasons this book, published in German only, was not noticed by the scientific community.

The lifelong *interdisciplinary* perspectives of Westenhöfer can also explain why some facets of his career have been marginalized, compared with others. There are reasons to emphasize that he was originally a pupil of the famous Professor Rudolf Virchow, at the University of Berlin, who transmitted his passion for both pathology and public health. There are also reasons to focus on what he learned as a doctor in the German army, particularly during World War 1.

There are also good reasons to focus on the Chilean phases of his life. Not only was he familiar with a great diversity of scientific disciplines but, furthermore, he was bicultural. One cannot understand who Westenhöfer was without referring to his first, second and third tenure in Chile.

During the first tenure (1908-1911), while he was hired by the government of Chile to teach Pathology at the School of Medicine of the University of Chile, he published a social medicine report where he described in critical terms the poor health conditions and hygiene practices that he observed in certain urban areas and nursing homes in Santiago. His paper led the government to deport him from Chile. In August 1911 there was a massive march in Santiago to protest at this expulsion. This is how the name of Westenhöfer became well known among some Chilean circles. During the second tenure (1929-1932) he directed the Department of Pathology at the Medical School of the University of Chile. His stress on the social determinants of mortality and morbidity influenced a generation of students, including Salvador Allende, a medical-student activist and later President of Chile. This is also the time when, by comparing carriers from different ethnic origins, he demonstrated that syphilis existed in America before the European conquest. During his final tenure in Chile (1948-1957), he served as a pathology adviser. His pupils founded the Department of Pathology of the Pontifical Catholic University of Chile. His biography is an opportunity to underline that *interdisciplinarity* is often associated with lateral thinking.

Alister Hardy

Alister Hardy was known as a respected emeritus Professor of Marine Biology at Oxford University, knighted in 1957. At the beginning of March 1960, he was invited to speak to a group of marine biologists at the Brighton Sub-Aqua Club. The title of his presentation was: 'Aquatic man: past, present and future.' In such a context, Alister Hardy thought he could speak freely with colleagues about his long-suppressed idea about human nature. He did not know that there were journalists in the room.

Two days later, sensational articles were published in the *Sunday Express* and other Sunday newspapers about the Oxford professor who claims that 'Man is a sea ape.' The scientific community was stunned. Wilfred Le Gros Clark, of the Natural History Museum, called Alister Hardy and said: 'Never do that again!' Alister Hardy tried to defuse the effects of such articles and rushed to publish in *New Scientist* a first article titled 'Was man more aquatic in the past?', followed soon after by a second article titled 'Will man be more aquatic in the future?'.

In these articles Alister Hardy mentioned, in particular, the loss of hair as a characteristic of a number of aquatic mammals, the direction in which the hairs lie on different parts of the body, the layer of fat adherent to the skin, the great number of sweat glands, bipedalism, and also occasional particularities such as a webbing between the second and third toes.

Most obituaries and biographies of Alister Hardy are so unbalanced that it is difficult to realize what kind of person he was. We must give great importance to some little-known phases of his life. Let us mention, among many other examples, that he was fascinated by balloon navigation. In 1924, while preparing for the Discovery expedition, he flew between London and Oxford with Ernest Willows, the famous balloonist. His book *Weekend with Willows* is an ideal opportunity to explore the personality traits of Alister Hardy. Reading between the lines one can guess his interest for technical details of piloting and, at the same time, his enthusiasm. His description of taking off is a model of concise transmission of feelings: 'One really felt that the balloon was one's little universe and the earth something quite separate that we had left behind.' He claims that it is thanks to this flight that he established links between what he learnt from balloon navigation and what he knew about the differential drift of plankton at deepening oceanic levels. What a perfect illustration of *interdisciplinary* thinking! His interest in colour preferences of insects is unusual among marine biologists! He designed a study involving vacuum cleaners fitted with separate hoses, each of which ended in an artificial flower of a particular colour.

It is highly significant that in his obituary, written by the marine biologist Freddy Marshall, on behalf of the Royal Society, there are twelve pages about his investigations of religious experiences and 'the spiritual nature of Man' versus one page about 'The aquatic ape hypothesis.'[2]

Carl Ortwin Sauer

Carl Ortwin Sauer is the first person to place nutrition at the heart of the evolutionary debate. He was a giant. He was Professor of Geography at the University of California at Berkeley from 1923 until 1957. His name might be selected to illustrate the concept of *interdisciplinary* perspective. He has been presented as 'The dean of American historical geography'.[3] He has always been concerned with the role of human beings in changing, intentionally or not, the face of the earth. He was a pioneer in terms of ecological awareness. In books such as *Agricultural Origins and Dispersals* and articles such as 'The Morphology of Landscape' he was constantly dealing with the forces that relate human life to the web of plant and animal life sustaining it. The scope of Sauer's work included topics as diverse as the timing of man's arrival in the Americas, the geography of Indian populations, and the development of agriculture and native crops in the Americas.

When he published, in 1962, a paper titled 'Seashore—Primitive Home of Man?' he was so innovative and ahead of his time that his point of view regarding human evolution was marginalized.[4] In retrospect we'll dare to claim that it was the phase of his career we should focus on. After emphasizing that our enormous brains are our defining feature, he could conclude that coastal areas are ideal places to satisfy human nutritional needs. He gave a great importance to the need of iodine at a time when medical experts were just considering the use of iodized salt to prevent goitres.[5] It was not yet understood that all over the world human beings struggle to meet their need for iodine, unless they have regular access to seafood. He also suggested that the fatty acids provided by the seafood chain are important to feed the brain, at a time when it was not yet known that DHA—the very long-chain polyunsaturated fatty acid of the omega 3 family—is a primary structural component of the brain. Let us recall that DHA is abundant and preformed in the seafood chain only.

Elaine Morgan

The common points between *Elaine Morgan* and the other pioneers of 'The aquatic ape hypothesis' are striking: they all developed the art of bridging fragmented disciplines. Algis Kuliukas, on the one hand, and Daryl Leeworthy, on the other hand, took the initiative to reveal, through biographical books, the extent of the *interdisciplinarity* of Elaine.[6,7] She won an exhibition (scholarship) to Oxford University where she read English Literature at Lady Margaret Hall. She was a highly successful freelance writer for television. Her output embraced plays, documentaries, series, serials and adaptations. She participated in a nine-part biography of Lloyd George and an adaptation of Vera Brittain's *Testament of Youth* which gained her the Writers Award of the Royal Television Society.

While these common points are obvious, the most effective way to interpret the unique role of Elaine as a catalyst at the root of radically new ways to explore human nature is to focus on the differences.

The first difference is that, while the previous pioneers were men, Elaine was the mother of three children. Motherhood is undoubtedly a factor influencing the ways of thinking. Another difference is that, where the other pioneers are concerned, the study of a possible aquatic phase in the history of Homo was a one-off phase in their career. After the Second World War, Max Westenhöfer went back to Chile and concentrated on other topics of interest than the evolution of Homo. After his opportunities to talk and write about 'The aquatic ape hypothesis' in 1960, it is as if Alister Hardy was impatient to go back to his lifelong interest in spiritual phenomena and the roots of religions. He founded the 'Religious Experience Research Centre' and published such books as *The Divine Flame* and *The Spiritual Nature of Man*. As for Carl Ortwin Sauer, after writing in a discreet way his paper about 'Seashore—Primitive Home of Man?', he published books on topics as diverse as *The Early Spanish Main* and 'Northern mists'.

The case of Elaine Morgan is radically different. From the day when she became aware of the point of view of Alister

Hardy regarding human evolution, the 'aquatic ape hypothesis' remained her leitmotiv (in spite of her weekly short columns in *The Pensioner* in the last years of her life). It has been the central theme of all her books: *The Descent of Woman, The Aquatic Ape, The Scars of Evolution, The Descent of the Child, The Aquatic Ape Hypothesis* and *The Naked Darwinist*. It was the reason for the symposia and seminars she organized in Europe and the USA.

She invited me to participate in two seminars: one in Southampton (England) and one in San Rafael (California). These events were ideal opportunities to know her personally and to understand her sense of humour as one of the roots of her charisma. In San Rafael, for example, she answered a question by the editor of *Mind/body Bulletin* by explaining why she wrote *The Aquatic Ape*: 'since scientists disliked my previous book (*The Descent of Woman*) because it was a bestseller, I understood the need for a book that would be neither funny nor feminist'.

All these pioneers have now disappeared. What next?

8

Straight to the point

The so-called 'aquatic ape hypothesis' might no longer be a controversial theory on the day when it is commonplace to take as a point of departure the defining feature of Homo, which is a supersized brain. The bases of the previous perspectives and the focus on the points of view of pioneers may distract attention from what is essential. Is it enough to know about the specific nutritional needs of the enormous human brain, particularly its needs in iodine and specific fatty acids?

It should be enough

We might take for granted that our ancestors could reach a degree of encephalization beyond the level seen in the great apes when they had not to struggle to meet their needs in brain-specific nutrients.

The issue of iodine is easy to summarize. Iodine plays an important role in thyroid hormone production, and therefore in brain development and brain functions. It is highly significant that iodine is the only nutrient for which governments legislate supplementation. Today the global problem of iodine deficiency primarily affects people not regularly consuming fish or shellfish. In people not regularly consuming 'shore-based' foods (fish, shellfish, turtles, frogs, plants, etc.), iodine adequacy depends heavily on artificially iodized table salt.

If some of our ancestors had easy access to iodine-rich food, they were undoubtedly those living in coastal areas.

The issue of the fatty acid DHA is more complex. It is not enough to repeat that this very long-chain omega 3 fatty acid is a primary structural component of the brain. We must first wonder how powerful the human enzymatic system is to synthetize a molecule which is preformed and abundant in the seafood chain only. Today the answer is that the synthesis of long-chain polyunsaturates in adults is extremely limited. Humans have the enzymes to make DHA, but the majority of studies on human adults show that this pathway is extremely inefficient.[1,2,3] The conversion is rarely above 0.5%, which is low compared to such mammals as rats and mice.[4] Furthermore, the enzymatic system requires the help of several nutritional co-factors, notably zinc and iron. Insufficient intake of these nutrients compromises even further this already inefficient pathway, making it completely unreliable. Metal nutrients needed for brain development and function are more bioavailable from fish and shellfish than from plant-based diets where their absorption is impaired by phytates and other 'antinutrients'.

Studying the fluctuations of brain size in our species inspires questions about the effects on foetal development of maternal nutritional factors. These questions were the reasons for a study we pursued in a London hospital in the early 1990.[5]

The objective of our study was to evaluate the effects of a prenatal nutritional counselling programme about the possible benefits of increasing dietary intake of sea fish. Four-hundred and ninety-nine pregnant women, attending selected clinics for antenatal care at Whipps Cross Hospital, before 20 weeks gestation, were offered a 20-minute nutritional advice session. They were encouraged to increase the intake of oily sea fish. For each woman interviewed a corresponding control was established. The mean neonatal head circumference and the head circumference for gestational age were significantly greater in the study group.

Our study was replicated and enlarged (3115 participants) at New Cross Hospital, in Wolverhampton.[6] Intervention group babies had significantly greater mean ($p < 0.001$) and adjusted for gestational age and sex ($p < 0.001$) head circumference.

The same kind of question inspired a study by Michael Crawford and his team about the effects of specific fatty acid supplementation in pregnant women on brain volumes of the newborn infants evaluated by MRI scans.[7] It appeared that males born to the supplemented mothers had significantly larger total brain volumes, total grey matter, corpus callosum and cortical volumes compared to the placebo group. Not only does this study show that maternal supplementation in specific fatty acids enhances newborn infants' brain size but it also suggests differential sex sensitivity of foetal brains to pregnancy fatty acids status.

This overview of the specific nutritional needs of the brain should be enough to convince anyone that Homo has many characteristics of a member of the chimpanzee family adapted to coastal areas. In fact, we must take into account a deep-rooted widespread conditioning. In many academic circles it will probably remain usual to repeat that the nutritional perspective is 'not enough'.

It is not enough

This is why we need to add a list of human characteristics that are 'not enough' … if considered one by one by highly specialized experts.

Nakedness, a trait shared with sea mammals, is 'not enough' …

The direction in which the hairs lie on different parts of the body is 'not enough' …

A subcutaneous layer of fat, another trait shared with sea mammals, is 'not enough' …

Bipedalism is 'not enough' …

A heavy head, which is correlated with needs in brain-specific nutrients, is incompatible with all types of quadrupedalism. On the other hand, bipedal humans can even add weights on their head while walking. Normal humans may be considered 'obligate' bipeds because the alternatives are very uncomfortable and usually only resorted to when walking is impossible.

The human vernix caseosa is 'not enough' …

The skin of human newborn babies is covered with vernix caseosa (literally cheesy varnish). It was commonplace until recently to claim that only the skin of human foetuses and neonates is covered by vernix caseosa, the greasy white substance secreted by the baby's sebaceous glands during foetal life. In many cultures the vernix was denied any role and routinely wiped away. Don Bowen, a marine biologist from Nova Scotia, Canada, revealed that the pups of seals also have vernix. Interestingly, he noticed that harbour seals (*Phoca viticula*), which swim with their mothers within minutes of being born, have more vernix than other seals, which do not swim for at least 10 days. It has high viscosity, suggesting that its water must reside within a highly structured state that is conferred by the abundance of water-filled foetal corneocytes.[8] These foetal corneocytes act as 'cellular sponges' that prevent water from moving across the skin, whereas sebaceous lipids, including squalene, provide a hydrophobic barrier. Vernix is so rich in squalene that a measure of its concentration in amniotic fluid had been suggested as a test to detect the effects of post maturity, since vernix disappears in prolonged pregnancies.[9] It is noticeable that squalene is more abundant in the aquatic animal kingdom than among other animals. Vernix caseosa might be interpreted as a transitory protection against immersion in non-isotonic water. We should at least remember that vernix caseosa is a common point between Homo and harbour seals (and late-term sea lions)[10], while it is unknown among land mammals.[11]

However, it is not enough …

The consumption of the placenta by land mammals 'is not enough' …

It seems that in our species 'placentophagy', i.e., the consumption of the placenta by the mother immediately after birth, as is observed in numerous terrestrial mammals, has never been instinctive.[12] If it had been at any time in the history of humanity, we should find traces of this behaviour in myths, legends, and reports from preliterate and pre-agricultural societies. I know of women who had reached a very instinctive state of consciousness in the perinatal period, behaving as if 'on another planet', and overcoming a great part of their cultural conditioning. Yet none of them had ever expressed a tendency to bring the placenta toward her mouth.[13] Modern women who occasionally have eaten pieces of placenta were inspired by theories, such as the theory that it might prevent postnatal depression. Scientific interest in the placenta has recently inspired such theories leading to a form of human placentophagy based on rational considerations. For example, the discovery by Kristal of a placental substance that makes endorphins more effective ('Placental Opioid-Enhancing

Factor') could be seen as a justification for placentophagy in our species.[14] However, we should avoid the conclusion that eating placenta is an innate human behaviour.

Exploring placentophagy is important since all land mammals eat the placenta. If eating the placenta has never been instinctive among our ancestors, this would be another common point with sea mammals, including cetaceans and seals.

Interestingly, in this regard, camels are the exceptions among land mammals: they do not eat the placenta. Camels have another particularity among land mammals: like Homo and sea mammals they have kidneys with medullary pyramids. Since camels consume highly salty plants and drink the water of salty ponds, and since sea mammals also have easy access to hypertonic salty substances, one can suggest that placentophagy might be correlated with the urgent need in specific nutrients, particularly minerals, in the postpartum period. It is as if placentophagy and non-pyramidal renal medullas were features shared by mammals that do not have access to hypertonic salty substances after parturition. Can camels from the desert help us to accept our waterside origins?

Meanwhile, 'it is not enough' …

-*The sense of smell* of human beings is mysteriously weak. It is the same among whales. When they separated from hoofed mammals about 60 million years ago and migrated to water, their sense of smell nearly disappeared.[15] However, it is not enough …

- *Body temperature control through the loss of sweat* is not a costly mechanism if we think of the human being as a primate adapted to environments where water and minerals are available without restriction. However, it is not enough …

-*A low larynx*, which gives us the ability to breathe through our nose or our mouth, is an anatomical particularity shared with sea lions and dugongs. However, it is not enough ...

-*A prominent nose* is a feature shared with the proboscis monkeys, the most aquatic of the primates. The proboscis live in coastal wetlands in Borneo. They are bipedal. They are excellent distance swimmers. They are able to swim 20 metres under water. Their toes are webbed. However, it is not enough ...

-*The human vagina*, like that of sea mammals, is long and oblique, and is protected by a hymen. However, it is not enough ...

- *The human apolipoprotein E gene* has more similarities with the one of sea mammals than with the one of land mammals (including common chimpanzees and bonobos).[16] Apolipoprotein E is the principal cholesterol carrier in the brain. However, it is not enough ...

-*Growth of bone in the ear canal* is particular to modern humans who are frequent swimmers, and also seals, as mammals adapted to both underwater life and life in the dry land. 'External auditory exostoses' are dense bony growths protruding in the ear canal at two or three constant sites. It is notable that they were first described in 1911 by Marcellin Boule in his classic monograph on the Neanderthal skeleton from the Bouffia Bonneval in La Chapelle-aux-Saints, France.[17] Since that time, it has been confirmed that the Neanderthals exhibit an exceptionally high level of auditory exostosis, in the region of 50%.[18] We must keep in mind that Neanderthals had enormous brains and, probably, a special relationship with water, since they could be skilled long-distance navigators. Today the colloquial term for these exostoses is 'surfer's ear' because, among modern humans, it is usually associated with frequent swimming and water immersion of the auditory canal. Many water sports have developed recently. This is a reason why 'surfer's ear' is becoming so topical in the specialized medical literature that some common beliefs are suddenly reconsidered. For example, a study assessed surfers living and surfing in Queensland, Australia.[19] It appeared that in young to quadragenarian-aged warm-water surfers, the prevalence and severity of auditory exostosis were much higher than anticipated: it had previously only been shown to occur in cold water surfers. This is in agreement with the report by Nicole Smith-Guzmán, who found seven cases of surfer's ears in males and one in a female skull in an ancient burial ground in a pre-Columbian village near the Gulf of Panama, at an equatorial latitude.[20] In such a renewed scientific context, Peter Rhys-Evans suggested that aural exostoses were evolved at a certain phase of the history of mankind for protection of the delicate tympanic membrane during swimming and diving by narrowing the ear canal in a similar fashion to other semiaquatic species.[21] It is interesting, but it is not enough …

-One of the most common abnormalities (or particularities) among humans is a *webbing between the second and the third toe*. When a congenital abnormality is an addition, it usually means that the feature was there for a reason during the evolutionary process. However, it is not enough …

-*A narrowing of the thoracic aorta* ('coarctation of the aorta') is common among humans and seals. However, it is not enough …

-*Menopause*, and prolonged life after reproduction, is a feature shared by humans, killer whales and short-finned pilot whales. However, it is not enough …

How many times shall we repeat 'it is not enough …' before daring to go one step further?

One step further

In the age of genetics, the best way to go one step further is to restart from the central event that separated the emerging *Homo* from *Pan* (the chimpanzees). We now have reasons to suggest that this event is at the root of the defining feature of Homo, which is the capacity to develop a large brain. The central event was the fusion of two chromosomes that remained separate among the other apes. Since this fusion, humans have only 23 pairs, while other primates have 24 pairs. More precisely the human chromosome 2 is an end-to-end fusion of chromosomes 2a and 2b. The divergence age between Homo and Pan remains imprecise because it has varied with the method of investigation. Before phrasing the useful questions, it helps to recall that chromosome 2 is the second-largest human chromosome, representing almost eight per cent of the total DNA in human cells, and that some chromosome 2 genes control the metabolism of fatty acids at different levels.

In the current scientific context, the point is to combine the two main particularities of Homo, provided by different perspectives. First, Homo has 23 pairs of chromosomes, with a large chromosome 2. Secondly, among the 260 species of living primates, only Homo has got the capacity to develop a large brain. Starting from these undisputable facts, the first question is about the kind of environment that might have facilitated the success of the crucial genetic mutation. A coastal environment seems probable, since the capacity to develop a large brain is associated with an increased need for brain-selective nutrients easily provided by seafood. A place that has been or has not been occasionally an island, according to the fluctuations of the sea levels, should be classified as ideal. Allan Krill has judiciously recalled that evolution of a new species by isolation on an island is comparatively much faster than elsewhere.

After the crucial genetic event that has distinguished our species from the other apes, adaptation to new environments probably facilitated other mutations, particularly among genes that play important roles in brain development, such as MCPH1.[22] Current evidence suggests that, beside alterations in the DNA sequence, modified gene-expression of *MCPH1* may also contribute to human brain development and function. During postnatal brain development, *MCPH1* is abundantly expressed in humans only. Recently, Chinese scientists reported inserting the human brain-related MCPH1 gene into Rhesus monkeys: the transgenic monkeys performed better and answered faster on short-term memory tests involving matching colours and shapes, compared to control monkeys. Brain-image and tissue-section analyses indicated a delayed neuronal maturation and myelination of the transgenic monkeys, similar to the known evolutionary change of developmental delay (neoteny) in humans.[23]

Probable genetic mutations and alterations of gene-expression explain the significant differences between humans in terms of brain size. For example, Homo erectus, the first human ancestor to spread throughout the world, having a distribution in Eurasia extending from the Iberian Peninsula to Java, had a 1,000 cc brain, while the brain of Neanderthals averaged more than 1,400 cc. As for Sapiens, it is around 1,350 cc.

Our objective is not to anticipate the outcomes of the generation of studies we are looking forward to but rather to focus on the new bases we now have at our disposal to clarify mysterious aspects of human nature.

9

The mammal that cannot swim

There are two main reasons why members of our species are different from other mammals:

- – A gigantic brain.
- – The need to learn a technique to be able to swim.

What are the links between these two particularities of human beings?

There are more than 5,000 known species of mammal in the world. Some of them have never been studied with regard to their swimming behaviour. Others, like giraffes, have well known particularities. Our point of departure is the general rule: mammals can swim.

When trying to interpret mysterious facts, the most promising attitude is usually to rephrase questions.

Some people have assumed that swimming was known in prehistorical times and wonder when the capacity to swim disappeared.[1]

Others have assumed that swimming is a modern habit and earlier in history people couldn't really swim for some reason. They wonder when humans started to swim.[2]

Until now, the questions raised by theoreticians were mostly related to the history of mankind. We suggest that the best way to go one step further is to focus on the history of individuals. It is to consider swimming behaviour at different phases of brain development. This perspective has been ignored until recently.

Learning from babies

The turning point came in 1939 when Myrtle McGraw published her article 'swimming behavior of the human infant' in the *Journal of Pediatrics*.[3] She filmed and wrote about 42 different babies during the two and a half first years of their life. She repeated observations of the same babies at different intervals, making a total of 445 observations in all. This enormous accumulation of data laid the foundations of a new understanding of aquatic behaviour in humans.

During the first phase of development, which concerns the first weeks of life and lasts no more than four months, the movements of the baby are striking when he/she is placed face down in water. There are rhythmical movements of flexion and extension in the legs and arms, together with swinging movements of the trunk from side to side, which propel the baby through the water. These movements are better organized and more pronounced when the baby is entirely submerged. This is an outstanding feature which arises from this pioneering study: the human newborn baby is perfectly adapted to immersion and automatically holds his/her breath when under water. The newborn baby looks happy, keeps its eyes open, and does not cough or show distress. The swimming movements and the control of breathing are two behaviours which tend to reinforce each other: when the baby is in the same position in water but supported under the chin, the swimming movements are not well organized.

After babies are four months or older, the movements become disorganized. Often the babies are quite inactive when supported under the chin and, when submerged tummy down, they tend to rotate on their backs. Then they appear to struggle, as if trying to clutch the hand of an adult. They sometimes swallow large amounts of water during that stage and are often coughing when their head emerges from the water. They need to be rescued.

The work of Myrtle McGraw is of paramount importance. She has provided keys to understanding human nature in general and some physiological functions in particular.[4] She has understood the concept of 'neocortical inhibition'. She has revealed that the capacity to swim is lost when the neocortex (the highly developed part of the human brain) is reaching a certain degree of maturation, obscuring functions that are under the control of primitive mammalian brain structures. We'll have opportunities to present the concept of 'neocortical inhibitory system' as a key to rediscover the basic needs of a labouring woman.

Learning from non-human apes

From the work of Myrtle McGraw we understand that human beings lose their innate capacity to swim when they reach a certain degree of neocortical maturity that takes place around the age of three or four months. This might be a key to interpreting conflicting data about the swimming behaviour of non-human apes. The first obvious difference between human beings and other apes is that human beings are usually attracted by aquatic environments. It is the opposite for non-human apes, to such a point that it has been commonplace, in zoos, to use water moats as limits to the territories of chimpanzees, gorillas and orangutans.

Today we have at our disposal well documented cases of swimming and diving apes. The most valuable ones are video-based studies of a swimming chimpanzee and a swimming orangutan by Renato Bender and Nicole Bender.[5]

Renato Bender was surprised when the chimpanzee Cooper dived repeatedly into a swimming pool and seemed to be very comfortable. To prevent the chimpanzee from drowning, the researchers stretched two ropes over the deepest part of the pool. Cooper became immediately interested in the ropes and, after a few minutes, he started diving into the two-metre-deep water to pick up objects on the bottom of the pool. Some weeks later, Cooper began to swim on the surface of the water. The orangutan Suryia also possesses this rare swimming and diving ability. Suryia can swim freely up to 12 metres.

Both animals use a leg movement similar to the human breaststroke 'frog kick'. While Cooper moves the hind legs synchronous, Suryia moves them alternatively.

The swimming behaviour of non-human apes is a complex issue. It has not been studied in depth until now. We have a lot to learn by comparing the power of neocortical control among human babies and among non-human apes.

From Plato to the next Olympic Games

Since time immemorial it has been understood that human beings cannot swim unless they have been taught a technique. The capacity to swim has often been considered in the framework of education. It has usually been a privilege associated with a positive connotation. Plato referred to uneducated people as those 'who can neither swim nor read or write...'.

Not only ancient Greeks, but also, for example, Egyptians, Chinese, Germans, Hungarians, Assyrians and people in Israel were learning the art of swimming. Herod (73 BC) made it obligatory for boys. In Japan learning to swim was an important part of training samurais. This was one of the noblemen's skills. In general, it seems that originally the capacity to swim has played an important role as a status symbol.

Depending on the cultural and historical contexts there have been several other reasons for acquiring the capacity to swim. In the Renaissance, after the non-swimming habit of the Middle Ages, books were published which covered the topic of how to save someone from drowning. The author of one of these books, the Dutch writer Fredéric Bachstrom, fought for swimming to be taught at school. At the beginning of the eighteenth century, the first known lifesaving group 'Chinklang Association for the Saving of Life' was established in China.

We must also keep in mind the motivations of warriors. In Japan, swimming was a part of the samurai's training over the millennia. The Japanese warriors developed a variety of swimming skills because Japan is surrounded by water where combat took place. In the army of Napoleon, swimming was obligatory. In the Napoleonic Wars, the number of drowned Habsburg soldiers, who could not swim, was disastrously high.

Swimming as a sport is a recent phenomenon. The Olympic Games in Athens in 1896 were a turning point. This is an opportunity to emphasize that, until the twentieth century, only men could be swimmers. It is significant that women were first allowed to swim in the Olympic Games in 1912 in Stockholm. It is also significant that Annette Kellerman, from Australia, was arrested in the USA in 1907 for indecent behaviour. She was practising a kind of synchronized swimming, diving into glass tanks… her swimsuit showed arms, legs and the neck.

Since swimming became a sport, the techniques have been continuously improving. Let us wait for the next Olympic Games …

Breaststroke versus front crawl: the realm of mysteries

Even after clarifying the reasons why adult human beings need to learn a technique to be able to swim, we still remain in the realm of mysteries. When considering non-human mammals, not only is there no need for learning and training, but also all members of one particular species swim exactly in the same way. Among humans, old techniques are constantly modified and new techniques appear. It seems that, for time immemorial, there have been two dominant families of swimming styles. In Europe, variants of breaststroke (including occasionally sidestroke) have been dominant until recently. For example, in 1875, Captain Matthew Webb crossed the English Channel, swimming the entire way breaststroke. In the Pacific Rim, variants of front crawl have been dominant.

In Europe, the front crawl was first seen in a swimming race held in 1844 in London where it was demonstrated by two Native Americans. They were invited by the British Swimming Society to race one another in an exhibition. After the demonstration, the English Society considered this style to be barbarically 'un-European' due to the splashing and kicking and the British continued to swim only the breaststroke in competition. During the following decades the techniques of front crawl were constantly modified, in particular by John Trudgen (the 'trudgen stroke') and by Alick Wickham, from the Solomon Islands, who introduced a highly effective modern method for fast swimming.

Advantages of learned behaviours

The reasons why human beings have developed two radically different dominant types of swimming styles may be considered mysterious. According to plausible interpretations, breaststroke was developed over time by people who swam in relatively tranquil conditions, such as ponds, lakes, rivers, and inland seas, while variants of front crawl were more useful in turbulent conditions, such as the big waves of the Pacific Ocean.

In fact, this great diversity of styles is first a reason to assume that, beyond a certain degree of brain development, swimming must be classified among learned behaviours: it occurs only after experience or practice. A learned behaviour is an advantage in terms of adaptability: it is flexible.

The time when our ancestors started to invent and develop swimming techniques has also been considered mysterious. Let us rephrase the usual questions and suggest that human nature probably includes the capacity to invent and develop swimming techniques. In other words, Homo has an innate capacity to learn to swim. There is no doubt that Palaeolithic Homos, particularly Neanderthals, were swimmers. This is the only way to explain their exceptionally high level of auditory exostosis, in the region of 50%.[6] Data from Grotta del Cavallo in southern Italy suggest that the exploitation of submerged aquatic resources was part of Neandertal behaviour well before the arrival of modern humans in Western Europe.[7]

In a renewed scientific context, swimming is becoming an inevitable issue for all interdisciplinary students in human nature.

Why is human birth occasionally easy?

Until now, a standardized question has continuously been raised: 'Why are human births difficult?' The responses were also standardized:

'It is for mechanical reasons.'

Such replies have often been justified by pictures comparing the size and shape of the maternal pelvis with the size and shape of the foetal head. There has been no need for long paragraphs since it is impossible to modify the bony pelvis of a pregnant woman. This is why, in textbooks, chapters about birth physiology among humans are short.

It will be another matter on the day when it is commonplace to wonder: 'why are human births occasionally easy?' Even during the twenty-first century, we all have heard about women who, surprisingly, gave birth very easily and quickly. Such anecdotes suggest that, until now, there has been a tendency to overestimate the importance of mechanical factors.

When the tool becomes the master

Once more, we'll first recall that a gigantic brain and a highly developed neocortex is what makes members of our species different from all other mammals. After interpreting swimming behaviours at different phases of human life we are in a position to understand that the new brain does not always play the role of a tool at the service of physiological functions. It can be the opposite. When a human behaviour or a human physiological function has mysterious characteristics, we must keep in mind that the activity of the powerful new brain (the neocortex) can have inhibitory effects: 'the tool can become the master'. We have a lot to learn from multiple examples of human physiological functions that are easily obscured by neocortical activity. To keep in mind the complexity of these issues, we'll just recall that there are both excitatory and inhibitory neocortical neurons.

The sense of smell, apparently weak among adult humans, is a typical example. An ingenious sophisticated experiment has eloquently demonstrated that the human sense of smell is improved following alcohol consumption: it is well known that alcohol reduces inhibition.[1] It is highly significant that olfaction appears as an important physiological function among newborn babies until a certain degree of neocortical development. I highlighted as early as the 1970s that, immediately after birth, it is mostly through the sense of smell that the baby is guided towards the nipple.[2,3] Since that time, there has been an intense curiosity, in scientific circles, for the functions of the sense of smell in early infancy.[4 to12] One of the reasons for this curiosity is that the sense of smell becomes gradually weaker after the first weeks following birth. The concept of neocortical inhibition offers the most valuable interpretation.

It is also worth recalling that in all human societies, even those where genital sexuality is comparatively free, couples isolate themselves to make love. Such a universally accepted need for privacy in specific situations indicates a deep-rooted understanding of an essential aspect of human nature.

Until now, the concept of neocortical inhibition is not commonly mentioned by medical practitioners, apart from psychiatrists: in the particular case of pathological conditions associated with culturally unacceptable behaviours, there is evidence for impaired inhibition. However, without being mentioned, the concept is understood in an empiric way by some clinicians. For example, a urologist specialized in prostatic diseases would never ask a man to urinate in front of him to evaluate the power of his urine jet.

It is artificial to separate the issue of neocortical inhibition from the issue of 'primitive reflexes'. These reflex actions are exhibited by normal infants, but disappear after some weeks or months and are not exhibited by normal adults. The 'rooting reflex', thanks to which a newborn baby can find the breast during the hour following birth, is usually mentioned in this framework. From a theoretical perspective, the important point is that older children and adults with pathological conditions such as cerebral palsy may retain these reflexes. Furthermore, such reflexes may reappear in neurological conditions such as dementia, strokes, and after traumatic lesions. It is also significant that primitive reflexes can reappear among normal elderly people: this is an obvious sign of the physiological ageing of the neocortex and its declining inhibitory power.[13]

When the concept of neocortical inhibition is assimilated, those who wonder why human births are occasionally easy will be ready to understand that reduced neocortical control is nature's solution.

The solution found by nature

Even during the twenty-first century, there are still people who know that when a woman is giving birth easily, there comes a time when she appears to cut herself off from our world. She forgets what she has been taught, what she read in books, and what her plans were. She can behave in a way that would usually be considered unbecoming to 'civilized' women, for example screaming or swearing. She can complain about odours nobody else can perceive. She can find herself in the most unexpected and bizarre primitive postures, often quadrupedal. A reduced neocortical activity can explain the well-known fact that many women forget details of what happened when they were in labour. Hundreds of women were interviewed about 10 days after giving birth. Those who had given birth by caesarean had a comparatively good recollection of many details.[14]

In a renewed scientific context, this is how we can summarize the basic needs of a labouring woman: she needs to feel protected against all possible neocortical stimulations. The keyword is 'protection'. Since language is a powerful neocortical stimulant, it is easy to reach the conclusion that silence is a basic need. One of the effects of the socialization of childbirth, as an aspect of the domination of nature that started about ten thousand years ago, has been to increase the risks of being exposed to language during labour. Of course, there are differences between several kinds of languages in terms of neocortical stimulation. There are differences between a language absorbed during foetal life and infancy versus a language learned as a university student. We may contrast baby language and intellectualized language. Asking a question is always a powerful way to stimulate neocortical activity. I have suggested that 'in the land of Utopia' the capacity to remain silent would be a prime selection criterion to be accepted as a midwifery student.

The effects of light on the birth process have not been taken seriously until recently, when it appeared that melatonin, the 'darkness hormone', is an essential birth hormone. This is why it is worth focusing on this factor. Studies of the interactions between melatonin and other brain mediators offer a promising avenue for research. The effects of melatonin as an inhibitor of neocortical activity are already well understood.[15,16] We feel obliged to deviate from the concept of neocortical inhibition and refer to recent advances regarding peripheral effects of melatonin. It is now established that there are melatonin receptors in the human uterus, and that melatonin works together with oxytocin to enhance contractility of uterine muscle. [17,18,19,20,21]

The point is that melatonin appears today as an important hormonal agent in human birth. This is confirmed by the significant amount of melatonin in the blood of neonates, except those born by pre-labour caesareans.[22] The importance of these findings appears clearly when the protective anti-oxidative properties of melatonin are taken into account. In the age of artificial lights, the reasons to improve our understanding of melatonin release and melatonin properties become obvious. We must realize that we are at a turning point in the history of light. Until now the emission of light was based on the phenomenon of incandescence: sun, fire, classical electrical bulbs, etc. Today light-emitting diode ('LED') is a semiconductor source emitting a light comparatively rich in the blue spectrum. It is already well established that this part of the spectrum is the most melatonin suppressive. The need to illuminate a birthing place is an effect of the socialization of the event.

As a general rule, we just need to keep in mind that all attention-enhancing situations stimulate neocortical activity and therefore inhibit the birth process. Feeling observed is a typical example of this kind of situation: in other words, privacy appears as a basic need. The perception of a possible danger is another typical situation: in other words, to feel secure also appears as a basic need.

After such considerations about what can make a human birth possible and even occasionally easy, we are in a position to describe and interpret the phase of preparation to what we have called 'the foetus ejection reflex'. The labouring woman may suddenly talk nonsense, as if she were on another planet. After thousands of years of socialization of childbirth and cultural conditioning, it is rare that an authentic foetus ejection reflex can follow. However, it is possible.

The foetus ejection reflex

I am often asked to clarify the difference between the foetus ejection reflex and the Ferguson reflex. Around 1940, when working with anesthetized rabbits, Ferguson had studied uterine contractions induced by vaginal dilation.[23] It was demonstrated afterwards that in the parturient sheep the 'Ferguson reflex' is associated with increased blood concentrations of oxytocin and utero-ovarian venous prostaglandins F levels.[24] It was also demonstrated that these responses are blocked by epidural anesthesia.[25] The results of such studies, focusing on vaginal dilation, have immediately attracted the attention of medical circles.

Niles Newton, on the other hand, looked at the effects of environmental factors on the birth of mice. By focusing on the importance of cortical inhibition, even among non-human mammals, she was studying parturition as a chapter of brain physiology.[26] She was using the term 'foetus ejection reflex'. It is significant that, compared with the work of Ferguson, the studies by Niles Newton have not attracted the same attention within obstetrical circles.

In the 1980s, I suggested that an authentic foetus ejection reflex is also possible among humans, but is usually repressed by neocortical activity. With the support of Niles Newton, I wrote that saving this term from oblivion would be a key to facilitating a radically new understanding of the particularities of human parturition.[27,28] I had observed that, in exceptionally rare situations, women do experience this reflex after a short series of irresistible and powerful contractions that precede the baby's birth. In such situations, voluntary movement has no place. When a typical reflex occurs, there is an obvious elimination of neocortical control: women seem to be 'on another planet', talking nonsense, behaving in a way usually considered unacceptable regarding civilized women, finding themselves in the most unexpected, bizarre, often mammalian bending forward or quadrupedal postures. The reflex does not always start at the same phase of descent of the foetal presenting part. A real foetus ejection reflex can occur long before the descent of the presenting part, or long after. It can start before complete dilation or after complete dilation. At the very time of the birth, mothers are typically in an ecstatic state; they may need some time to realize that the baby is born and dare to take it in their arms.

It is easy to explain why the concept of foetus ejection reflex is not understood after thousands of years of socialization of childbirth. It is precisely when delivery seems to be imminent that the birth attendant tends to become even more intrusive. The foetus ejection reflex can be preceded by sudden, explosive expression of a fear with a frequent reference to death.[29] Any attempt to reassure with words can interrupt the progress towards the foetus ejection reflex.[30] In the particular context of a preliterate and pre-agricultural society of New Guinea, Wulf Schiefenhovel could discreetly film women giving birth in the bush, without any assistance, through authentic foetus ejection reflexes similar to those some modern women can occasionally experience in ideal situations.[31] In general, any interference tends to bring the labouring woman 'back down to Earth' and tends to transform the foetus ejection reflex into a second stage of labour which involves voluntary movements.

Today, there is no simple recipe to overcome thousands of years of cultural conditioning. However, I can describe an environment, at the limit of utopia in the case of civilized modern women, that is compatible with an authentic foetus ejection reflex. The reflex is more likely to occur in a small dark room at a comfortable ambient temperature, with no one around.

The popularization of the concept of foetus ejection reflex, as a consequence of a renewed understanding of birth physiology, would be a critical step towards a certain degree of desocialization of childbirth. Can we present authentic midwifery as the art of *protecting* an environment compatible with a foetus ejection reflex?

Challenging thousands of years of tradition

By presenting childbirth in the light of modern physiology, we are challenging thousands of years of cultural conditioning. We present the birth process as an involuntary process under the control of primitive brain structures we share with the other mammals. As a general rule, one cannot help an involuntary process, but one can identify inhibitory factors. From a practical point of view, in terms of basic needs, the key word is 'protection'. Since our point of departure was the concept of neocortical inhibition among humans, protection against language was given first place. If our point of departure had been the concept of adrenaline-oxytocin antagonism among mammals in general, we would have first mentioned protection from scary situations and low ambient temperature to explain that when mammals in general release emergency hormones of the adrenaline family, they cannot release oxytocin, the main birth hormone. As a general rule, in emergency situations, mammals postpone physiological processes related to reproduction when they need the energy to fight or run away.

This reference to the adrenaline-oxytocin antagonism is an opportunity to recall that in the middle of the twentieth century, when I was a medical student ('un externe') in the department of obstetrics of a Paris hospital, midwives were spending their life knitting.

At the April 2004 British Psychological Society conference, Emily Holmes presented her studies of the effects of repetitive tasks, such as knitting, in stressful situations.[32] The author could conclude from her studies that repetitive tasks are extremely effective means of reducing tensions. Emily Holmes underlined that her research was consistent with the actions of notorious French tricoteuses of the French Revolution, such as Madame Defarges, who knitted while they watched people being guillotined, never, apparently, experiencing post-traumatic stress disorder. She also referred to the use of worry beads in many cultures, such as Greece, as a way to cope with stressful situations.

We might translate such findings in physiological language and realize that when midwives spend long hours knitting, their own levels of adrenaline are kept as low as possible. Since high levels of adrenaline are extremely contagious, the progress of labour is to a great extent dependent on the level of adrenaline of those who, for sociocultural reasons, might feel obliged to be around.[33,34] What is the future of knitting sessions in midwifery schools?

To present 'protection' as a keyword is a preliminary, simple and concise way to challenge tradition. Since the beginning of the socialization of childbirth, as an aspect of the domination of nature associated with the 'Neolithic revolution', the basis of our cultural conditioning is that a woman has not the power to give birth by herself. She needs some kinds of cultural interferences. The dominant paradigm has undergone many phases, from the advent of midwifery and the most deep-rooted perinatal beliefs and rituals up to the current masculinization and medicalization of the birth environment. The current key words are eminently disempowering. They always focus on the active role of somebody else than mother and baby, the two obligatory actors in the birth drama. They are variants of the concepts of helping, guiding and controlling. The terms 'coaching', used by groups promoting 'natural childbirth', and 'labour management', used in medical circles, imply the intervention of an expert, while the term 'support' suggests that to give birth a woman needs energy brought by somebody else.

On the day when the term 'protection' overrides the current watchwords, we will have reached a turning point in our understanding of human birth.

11

The Oceanic feeling

One cannot imagine a study of Homo and the 'Planet Ocean' without a reference to the 'oceanic feeling', as a key to understanding human nature.

It is unusual to claim that swimming, giving birth and access to transcendence are facets of the same topic. However, the links are easily perceived by anyone who has assimilated the concept of neocortical inhibition.

The highways to transcendence

We are now in a position to interpret what Romain Rolland, in his correspondence with Freud, called 'the oceanic feeling',[1] and what I call, in the twenty-first century scientific context, 'transcendent emotional states'.[2] Both 'oceanic' and 'transcendent' are adjectives that suggest the absence of limits.

Once more our point of departure is to consider different phases of individual development. This is why, as we did for swimming, we'll first wonder what we can learn from babies. Since Freud, it is well accepted that there is a 'primitive ego-feeling' in infancy: a baby cannot differentiate himself or herself from other people and things. After that, it is as if human beings, after a certain phase of brain development, gradually lose their capacity to escape from space and time reality, except in special occasions.

These special occasions are not easily studied today because, since the beginning of the domination of nature by our ancestors, they have been regulated and channelled by cultural milieus. We must keep in mind that the domination of nature, that started about 10,000 years ago, included not only the domestication of plants and animals, but also the domestication of human beings. All societies have developed, promoted and given directions to paths towards transcendence that are culturally acceptable. For example, prayer, fasting, music, dances, songs and psychedelic drugs may be culturally controlled routes to transcendence. Modern scientific research is influenced by the dominant deep-rooted cultural conditioning: in general, only states of consciousness deliberately induced by humans are widely studied.

On the other hand, ecstatic/orgasmic states related to reproduction are not easily accepted topics in academic circles. They are usually associated with negative connotations. An analysis of the vocabulary is an eloquent way to realize that organs and physiological functions directly involved in human reproduction are associated with the concept of shame. In English, for example, 'pudenda' is used as the scientific term for external genital organs, which are innerved by the 'pudendal nerves' and receive their blood via the 'pudendal arteries'. The root of these terms is the Latin verb 'pudere', which means 'to be ashamed'. It is the same in a great diversity of languages. In Chinese the pubic bone is called 'Chigu', which literally means 'shame bone'. In Japanese the penis is often called 'the son', suggesting that the use of a specific or too explicit term would be a violation of the norms of the culture that would induce a feeling of shame.

It is as if, for thousands of years, many aspects of human nature have been concealed by a deep-rooted cultural conditioning. In a renewed scientific context, we cannot ignore that certain physiological functions, which have routinely been altered for ages, are potentially associated with ecstatic states. This is how I have introduced the concept of 'highways to transcendence' when referring to episodes of human reproductive life such as the 'foetus ejection reflex', the 'milk ejection reflex' and the 'sperm ejection reflex'. All these events have similarities that might be studied in the framework of specific hormonal states associated with a reduced neocortical control. Without multiplying anecdotes, I'll just underline that the exceptionally rare modern women who gave birth through an authentic foetus ejection reflex and who were in an ecstatic state at the time of the very first contact with the newborn baby claim that they have been transformed by the experience.

There are reasons to suggest that elderly human beings, with weakening neocortical activity, might have an increased capacity to experience transcendent emotional states. This is still in the framework of empiric knowledge, although a new generation of scientific studies provides reasons to deepen this topic.[3]

Specific pathological conditions might also be classified as kinds of 'highways to transcendence'. This is the case of the 'Geschwind syndrome' in the framework of temporal lobe epilepsy.[4] This syndrome includes easy access to transcendent emotional states. It seems to be linked to a special connectivity between temporal neocortex and primitive brain structures.[5] Although it is difficult, in retrospect, to establish firm and uncontested diagnoses, it has been claimed that many well-known mystics had pathological episodes compatible with temporal epilepsy. It has been pointed out, for example, that during his conversion on the road to Damascus, Saint Paul had three days of blindness, fell down to the ground, and experienced ecstatic visions. Muhammad described falling episodes accompanied by visual and auditory hallucinations. Joseph Smith, who founded Mormonism, reported lapses of consciousness and speech arrest, and once found himself lying on the ground. As for Joan of Arc, a great light accompanied the voice she heard.

Half-open doors to transcendence

Enthusiasm is not included by scientists within their conventional lists of basic emotional states and emotional traits. As suggested by the roots of the word, an enthusiastic person is inspired by divinities. Enthusiasm is a way to transcend the objective daily reality. It can be presented as a variant of transcendent emotional states. Since the role of religions is to channel the universal transcendent emotional states, it is no wonder there have always been strong links between the concept of enthusiasm and the concept of religion. For example, 'The Enthusiasts' was the name of a Syrian religious sect of the fourth century; Protestant sects of the sixteenth and seventeenth centuries were called enthusiastic.

Keeping in mind that the routinely disturbed transition between intrauterine and extrauterine life occurs during a crucial period of the development of the emotional brain, we cannot help raise questions about the future, at a cultural scale, of transcendent emotional states in general. These questions are first and foremost about the origin of inspiration and therefore scientific and artistic creativity. Like enthusiasm and transcendent emotional states in general, inspiration is not frequently studied by modern psychology. The mysterious burst of creativity called inspiration has always been considered a divine spark. In Greece the muses were the Goddesses of inspiration in literature, science and the arts. Socrates taught that the inspiration of poets is a form of enthusiasm.

What would be the future of a hyper-brainy Homo deprived of enthusiasm and inspiration?

Joy can also be classified among the half-open doors to transcendence. Once I was asked to talk at a conference about 'the function of joy'. I had to overcome a major obstacle: although many emotional states have been studied in a scientific way by physiologists, psychologists, epidemiologists and other scientists, the concept of joy has not. Explore scientific and medical databases: the keywords 'anxiety', 'stress', 'depression', 'psychological distress', or 'fear' bring up thousands of references. 'Joy', on the other hand, remains a comparatively sterile keyword.

Since it has been said that painters, poets and other artists always precede scientists, I first looked at what we can learn from them about joy. 'The Five Joyful Mysteries' is undoubtedly the most fruitful reference in the field of painting and it is significant that the word 'joy' is associated with the word 'mystery' because 'mystery' has the same root as 'mystic'. In other words, a joyful experience transcends the limits of ordinary experience and takes us into the realms of the mystic. The Annunciation, the Visitation, the Nativity, the Presentation in the Temple and the Finding in the Temple are all events related to the emergence of life. Archetypal joyful experiences are related to maternal love. They are intense responses to rewarding events. We also learn that joy is contagious. After the Annunciation, Mary will share her joy with another mother-to-be. The sacred atmosphere of the Temple is appropriate for the expression of different facets of maternal love.

Poets and musicians are not strangers to joy. 'The Ode to Joy' is now the European anthem, based on the fourth movement of Beethoven's ninth Symphony. One can wonder how the music by Beethoven evokes joy… sudden intermittent series of ascending notes are undoubtedly suggestive of the emergence of life. The original text of the European anthem was the poem written by Friedrich Schiller at the end of the eighteenth century. From the start of the poem, joy is presented as sudden access to the divine: 'Freude, schöner Götterfunken' (Joy, beautiful spark of divinity). The last line of a poem in alexandrines by my mother about joy ('joie') is also highly significant:

'*Un grand hymne à la joie évoque le Très-Haut*'[6],

which can be translated as 'a grand hymn of joy evokes the Almighty'. Poets also symbolically associate joy with the emergence of life. The poem by my mother includes the words '*printemps*' (spring), '*oiseau qui chante*' (singing bird), '*enfant*' (child).

12

Humanity and Mother Ocean

The concept of respect for 'Mother Earth' is ancient. It was a basic component of proto-Indo-European mythologies. In the age of ecological awareness, it has a sudden popular resurgence, strengthened by modern means of communication such as YouTube, songs, t- shirts, novels and videos.

From Green to Blue

Today we must express a warning regarding the possible effects of restrictive interpretations of the dominant vocabulary and the dominant symbols.

There is a risk that the word 'earth' makes many people visualize the continents. We must keep in mind that in a great diversity of languages there is only one word for 'earth' and 'land', although 70% of our planet is covered by water: in French, for example, '*la terre*' means either 'land' or 'planet Earth'. By renewing a vocabulary established when the Earth was not visualized as a globe, we might help modern humans to realize that 95 per cent of the planet's habitable space lies within the deep immense oceans: a sperm whale descends 1,000 metres to look for food. We must identify the priorities, recondition ourselves and think in terms of protection of the 'Mother Ocean'. Our planet is the only astronomical object known to have bodies of liquid water on its surface. The oceans may be presented as the givers and sustainers of life. We can make similar comments about the symbolism of colours. Green still symbolizes Nature, the natural world and ecological awareness. Why not blue, the colour of the oceans? Even in Great Britain, after the film-makers in the BBC's Natural History Unit presented *The Blue Planet*, narrated by David Attenborough, green politics has not yet been reduced to a tiny chapter of 'blue politics'.

Until now we have repetitively used the term 'Homo', referring to one particular species of primate. More often than not we have been thinking like zoologists. From now on, when considering the future of 'Mother Ocean' in relation to human activities, we need a grand-scale paradigm shift. We must think in terms of *humanity*.

To introduce the concept of humanity, the first step is to realize that human beings are unique among primates. All other primates are characterized by specific social structures and mating systems. We'll just focus on one example to illustrate this general rule. Hamadryas baboons have a well-documented four-level social system. Most social interaction occurs within small groups called one-male units or harems containing one male and up to ten females, which the males lead and guard. A harem often includes a younger 'follower' male that may be related to the leader. Two or more harems unite repeatedly to form clans. Bands and troops are the next levels. Social structures are also well documented among other social primates such as the common chimpanzees, the bonobos, the gorillas and the Rhesus macaques.

Regarding the big-brained Palaeolithic Homo, we cannot describe specific social structures and mating systems. The term 'tribe' is commonly used. We have the proof that interactions between groups, including interbreeding between varieties of Homo, were frequent, since a significant fraction of the genetic material of most of us comes from Neanderthal or Denisovan. After the advent of agriculture and animal husbandry (the 'Neolithic crisis') human groups became larger and larger. There have been villages, towns, cities, provinces, states, nations and international unions. We have reached the ultimate dimension with the concept of globalization. We are entering a new phase in the history of life.

How can humanity develop its respect for 'Planet Ocean'? This is a way to rephrase a question I raised (prematurely!) in 1979, in a book titled *Genese de l'homme écologique*.[1] At that time, I classified the respect for our planet in the framework of love of Nature and I raised questions about the development of the capacity to love. We have now realized how critical the period surrounding birth is and we have learned about the behavioural effects of oxytocin and other components of the hormonal flows released during the birth process. We are in a position to observe that, in the age of safe techniques of caesareans and cheap pharmacological assistance, love hormones are redundant, at a global scale, in this critical period. We must train ourselves to think in terms of humanity. For example, parents should be reassured about the fate of one particular baby born with pharmacological assistance or by caesarean section, because the cultural milieus can compensate many individual deprivations. From now on, the focal questions should be about the evolution of humanity. What is the future of humanity if nearly all members of our species are born without hormones of love? I cannot imagine more vital questions.

Disclosing priorities

Meanwhile, because our individual power is undoubtedly infinitesimal, the prerequisite to urgently initiate a collective awareness is to disclose priorities.[2,3]

At a time when the focus is on global warming, it is essential to know that coastal systems such as mangroves, salt marshes and seagrass meadows have the ability to absorb, or sequester carbon at rates up to 50 times those of the same area of tropical forest. Total carbon deposits in these coastal systems may be up to five times the carbon stored in tropical forests. We must recall that between 1980 and 2005, 35,000 square kilometres of mangroves were removed globally and that between 30 and 35 per cent of the global extent of critical marine habitats such as seagrasses, mangroves and coral reefs are estimated to have been destroyed. It is also becoming urgent to realize that carbon in faeces, respiration and other excretions from fishes—roughly 1.65 billion tons annually—make up about 16 per cent of the total carbon that sinks below the ocean's upper layers. This is one of the many urgent reasons to train ourselves to think long-term: carbon that makes its way below the sunlit layer becomes sequestered, or stored, in the ocean for hundreds of years or more. This key part of the Earth's biological pump will help scientists understand the impact of climate change and seafood harvesting on the role of fishes in the movement of carbon in the ocean, including from the surface to the deep sea.[4]

At the same time, we must keep in mind that a huge amount of carbon stored at the bottom of the ocean is released every year as massive nets are dragged along the seabed, whirling up marine sediment. Scientists estimate that CO_2 emissions from bottom trawling exceed carbon emissions from global air travel![5]

Ocean acidification is suddenly becoming a dominant preoccupation. It has been mysteriously overlooked until recently, despite decades of attention to rising levels of carbon dioxide in the atmosphere, and although it is easy to anticipate that when carbon dioxide dissolves in the ocean it lowers the pH, making the ocean more acidic. It was first highlighted in 2003 in a study by Ken Caldeira and Michael Wickett.[6] The authors quantified the changes in ocean pH that may result from a continued release of CO_2 and compare these with pH changes estimated from geological and historical records. They found that oceanic absorption of CO_2 from human activities may result in larger pH changes over the next several centuries than any inferred from the geological record of the past 300 million years, with the possible exception of those resulting from rare, extreme events such as bolide impacts or catastrophic methane hydrate degassing.

Ocean acidification may threaten phytoplankton, which forms the basis of the ocean food chain up to fish, marine mammals and ultimately human consumption. Ocean acidification may render most regions of the ocean inhospitable to coral reefs. Coral reefs are the nurseries of the oceans: they are biodiversity hot spots. On some tropical coral reefs, for example, there can be 1,000 species per m². We must add that an increase concentration of atmospheric CO_2 will make the ocean corrosive to the shells of many marine organisms.

It is probable that in the near future great importance will be given to the fast declining of the major ocean-circulation system which gives rise to the Gulf Stream. The slackening currents are likely to produce more extreme weather at places where the Gulf Stream normally brings a mild climate. According to the author of a recent study, it is likely to weaken further in 20 to 30 years, and that will inevitably influence our weather, so we would see an increase in storms and heatwaves in Europe, and sea level rises on the east coast of the USA.[7]

In this book about human nature, our objective is not to provide an exhaustive list of the multiple known effects of human activities, including overfishing, on the evolution of the oceans. It is more often than not difficult to transmit updated information and to select valuable predictions.

This is the case of mercury pollution. Fish and shellfish concentrate mercury, often in the form of methylmercury. Species of fish that are long-lived and high on the food chain contain higher concentrations of mercury than others. The differences between species are enormous. For example, the concentrations of mercury are about one hundred times higher among swordfish than among sardines. Mercury is undoubtedly toxic, especially due to its ability to damage the central nervous system. The presence of mercury in fish can be a particular health concern for women who are or may become pregnant, nursing mothers, and young children. Mercury release occurs by both natural and anthropogenic processes. It is difficult to make predictions, because the release by a great diversity of anthropogenic processes, such as fossil fuel burning, are fluctuating.

This is also the case of plastic pollution. Can we take for granted the predictions of scientists who claim that, by 2050, the quantity of plastics in the oceans will outweigh fish? Meanwhile everybody, including school children, must understand that plastic doesn't break down like other organic waste but turns into microplastics which is then easily mistaken by fish and invertebrates as plankton. The consumption of plastics by fish, caught for food for humans, allows plastics to travel up our food chain. Health consequences include ingesting the carcinogenic pollutants that attach to microplastics. The half-life of commonly discarded plastics is still being debated. Factors such as ocean surge and ultraviolet light will accelerate the breakdown of the original product; what remains are microplastics which can be readily consumed by plankton before working their way up the food chain. There is little doubt that plastic pollution has affected 100% of marine turtles, 59% of whales, 36% of seals and 40% of seabirds of those examined. Over 1 million seabirds and 100,000 marine mammals are killed by ocean plastic every year. Hundreds of species of marine animals are in danger of extinction due to plastic.

For some species, it is difficult to evaluate the comparative importance of several factors explaining their collapse. This is the case of sharks and rays, the 'Ocean's apex predators'. They have declined by 71 per cent since 1970. It is a very serious issue because they are at the end of the seafood chain. Their numbers are therefore indicators of the evolution of the oceans.

Beyond the explicit messages

Explicit messages based on a superficial knowledge of the spectacular recent transformations of 'Mother Ocean' should now be sufficient to identify priorities in the initiation of a new awareness. For example, most of us don't need a deep knowledge of modern irrational fishing techniques to understand how urgent it is to multiply 'Marine protected areas'.

However, we must not underestimate the possible complementary effects of subtle messages. This is exactly what Heloisa Lessa, a Brazilian midwife, and I had in mind when we organized, in February 2010, the 'Mid-Atlantic conference on birth and primal health research' in Las Palmas, Gran Canarias. Many of the 1,600 participants understood that through the title and the location of the event our objective was to suggest the links between the health of human beings and the evolution of the oceans. In October 2012, we sent a similar message through the 'Mid-Pacific conference on birth and primal health research' in Honolulu.

This toing and froing between explicit and subtle messages is an opportunity to recall that today scientific advances are moving fast. Many of us feel the need to be constantly updated. One way to satisfy this need is to rely on weekly, monthly or quarterly publications about the evolution of the marine ecosystems. Several of these publications provide valuable information to a general public, particularly on vital issues such as the regulation of fishing.

13

From a garden paddling pool to the Pacific Ocean

Jumping from one scale to another

Our relationship with space and time is an obligatory topic for all interdisciplinary students in human nature. Homo—the primate endowed with a powerful neocortical supercomputer—has an immense capacity to leap from one order of magnitude to another.

For example, we learn from some scientists that the diameter of an atom of helium is about 0.06 nanometres. A nanometre is one billionth of a metre (0.00000000 m). At the same time, we can learn that the distance from the Earth to the edge of the visible universe is about 45.7 billion light-years. A light-year is equivalent to about 9.46 trillion kilometres.

Where evaluation of time is concerned, we need millisecond as a unit when studying liquid-crystal displays (LCDs) that have a wide range of applications. A millisecond is a thousandth of a second. In contrast, we must be able to think in terms of hundreds of thousand years when considering the protection of an ecosystem against nuclear wastes, whilst we need to think in terms of billion years when referring to the history of life on Earth.

The turning point in the history of the relationship between Homo and space and time is symbolized by the work of Galileo, during the seventeenth century. The term 'microscope' was coined in 1625, when Galileo presented to the Accademia dei Lincei a new instrument that makes visible objects imperceptible to the unaided eye. At the same time Galileo, as the 'Father of observational astronomy', pointed a telescope skyward. He was able to discern mountains and craters on the moon, as well as a ribbon of diffuse light arching across the sky—the Milky Way. He also discovered the rings of Saturn, sunspots and four of Jupiter's moons.

Before this turning point, the ability to shift from one scale of magnitude to another had already appeared as an essential aspect of human nature. Objects resembling lenses date back 4,000 years and there are Greek accounts from the fifth century BC of the optical properties of water-filled spheres.

A 32,500-year-old carved ivory mammoth tusk could contain the oldest known star chart (resembling the constellation Orion).[1] Where the relationship with time is concerned, it is worth noticing that there were Sumerian, Egyptian, Assyrian and Elamite calendars.

Interdisciplinary perspectives imply the capacity to adapt to a great diversity of orders of magnitude. Our study of the relationship between Homo and water led us to introduce issues such as the colonization of the Pacific Rim by 'Homo navigator'. My dormant interest in 'Homo and Water' was awakened in the age of tap water with objects the size of a garden paddling pool.

In the age of tap water

I realized the power of a watery environment in the 1970s, at a time when one of my objectives was to develop strategies to reduce the need for pharmacological assistance in the particular case of women in established labour with unbearable back pain, when the dilation of the cervix could not progress.

In such a context I postulated that immersion in water at body temperature should also be a way to break a vicious circle by inducing a state of relaxation and lowering the levels of adrenaline. This is how, as a temporary measure, we had a big garden paddling pool. This was the beginning of the history of hospital birthing pools. Later on, we installed a larger, round, deep-blue bath that was plumbed in. On the walls of the small aquatic birthing room there were pictures of dolphins. I had a revelation the day a woman gave birth on the floor before the pool was actually full. All she needed was to see the blue water and hear the noise of water. Until that moment I had only simplistic physiological explanations regarding the effects of water immersion during labour.

This was a first lesson: when in labour, many women are irresistibly attracted to water and a great importance should be given to the way they are introduced in the aquatic birthing room.

The second important lesson was that, in general, when a woman enters the birthing pool in well-established labour, there is spectacular progress of the dilation of the cervix during the first two hours. The progress of cervical dilation is usually associated with behaviour suggestive of a remarkable reduction in neocortical control. Ideally, the bath should be deep enough for total immersion and the water temperature comfortable (in practice around 37° Celsius). A dim light is, of course, an important factor and, if it is acceptable, there would be nobody around apart from one silent, low profile and motherly birth attendant who does not behave as an observer. After two hours there is usually a feedback mechanism and the contractions become increasingly ineffective.

At the time of the garden inflatable pool (before we installed a proper plumbed-in bath), women were not influenced by the media or by what they read in books about childbirth. Their behaviour was spontaneous and thus we learned about the genuine effects of a water environment. A typical scenario (with many possible variations) was the case of a woman entering the pool in hard labour, spending an hour or two in water and then feeling the need to get out of the pool when the contractions were becoming less effective. This going back to dry land often induced a short series of irresistible and powerful contractions so that the baby was born within several minutes.

One day, a mother-to-be had not been in water for long when suddenly she had two irresistible contractions and the baby was born before she felt any need to get out of the pool. While giving birth, this woman was really 'on another planet'. Clearly, in that altered state of consciousness associated with hard labour, she intuitively knew that her baby could be born safely under water. There was no panic. It is as if a deep-rooted knowing was able to express itself as soon as the intellect and its knowledge were set aside. Such births happened again. It is as if, while in a particular state of consciousness, some women knew that a birth under water was safe for the baby.

When we had the experience of one hundred babies born under water (while thousands of women had used the birthing pool), I found it relevant to publish our observations in a mainstream medical journal.[2] This was an opportunity to warn my colleagues that in any hospital where a birthing pool will be available, a birth under water is bound to occur occasionally, even if it is not intentional. It is notable that this original message was distorted: many women were the prisoners of their project of 'water birth', remaining in the birthing pool when the contractions were becoming increasingly ineffective. For that reason, I tried—with limited success—to recall the original message in midwifery journals.[3,4,5]

Talented journalists usually have their fingers on the pulse of what interests the general public. They can anticipate how their audience (readers, viewers or listeners) will respond emotionally. They know, or rather they have an instinct, about what will capture the public's imagination. We had a lot to learn from the incredible interest shown by journalists in birth under water. At first, we were often irritated by reporters who came to our maternity unit and seemed interested in nothing else than birth under water. I tried unsuccessfully to attract their attention, in the context of a general hospital, on the small home-like birthing room and the singing sessions for pregnant women. I also tried to explain how we had learned that, in humans, lactation is supposed to start during the hour following birth. I learned from the discrepancy between what is 'mediatic' and what is essential.

In the late 1970s we were also visited several times by Jacques Mayol, the famous diver who became the hero of the film *The Big Blue*. Before we had heard about 'the aquatic ape hypothesis' through the books by Elaine Morgan, he was already convinced that Homo has special links with the sea and in particular with dolphins.

Once, during a conversation with Jacques Mayol, the name of Igor Tcharkovsky appeared: it had been associated, in the media, with birth under water in Moscow. I mentioned that I had already contacted the embassy of the Soviet Union in Paris to try to communicate with 'Dr Igor Tcharkovsky'. The embassy politely answered that they could not find this name in the list of Russian medical practitioners. My first interpretation was that this doctor was probably a legend born in the imagination of a clever journalist. Jacques Mayol reacted by mentioning that he had good friends in Moscow who might help him to clarify the enigma. Some days later I received a postcard from Moscow signed by Igor and Jacques!

This is how the enigma was solved. Igor Tcharkovsky was not a doctor, but a swimming instructor who had focused his interest on the capacities of human babies in water. He had realized that the younger the baby is, the better his adaptation to water. The climax of the 'Tcharkovsky phenomenon' was reached on the day when the media reported fascinating anecdotes of babies born in the Black Sea among dolphins!

Before the age of tap water

We can at least affirm that legends of aquatic births and dreams associating birth and water are not new. In ancient Greece, Aphrodite, the Goddess of Love, was born from the foam of the waves. As for the Cyprian Goddess of Love, she was born on the beach at Paphos. Freud interpreted dreams of water as being about birth. According to oral tradition, Japanese women living in some small villages by the sea used to give birth in water. Engravings suggest that in some African tribes the traditional place to give birth was near a river. Some Aborigines on the western coast of Australia used to first paddle in the sea before giving birth on the beach. Birth under water was probably known in cultures as diverse as the Indians of Panama and, perhaps, some Maoris of New Zealand. The only valuable written document is a book by the American linguist Daniel Everett, who spent a great part of his life among the Piraha, an ethnic group of Amazonians along the River Maici.[6] In the dry season, when there are beaches along the Maici, the most common form of childbirth is for the woman to go, usually alone, into the river up to her waist, then squat down so that the baby is born into the river. It is supposed to be cleaner and healthier for the baby.

Until recently, in our societies, nobody would have thought of considering the effects of a watery environment on the way women give birth. However, it is worth meditating over the widespread ritual of boiling water. Since this ritual preceded by far the Pasteur era, the rationale could not be to make the water sterile. Occasionally, it could have been a way to neutralize the invasive husband by keeping him busy. We can say for sure that it was a way to create a watery environment. It is worth noting that an interest for the issue of birth and water started to openly develop when everybody had easy access to cold and hot tap water. It is probable that, until that time, the attraction to water during labour could not express itself. It was simply obscured for practical reasons.

As usual, we'll conclude with questions. Is the current widespread use of birthing pools and the publication of countless books and articles about the use of water in childbirth the expression of a transitory fad? Or, on the contrary, are there reasons to think that it is the expression of a deep-rooted aspect of human nature?[7]

References

Chapter 1

1 Michel Odent. *Water, Birth and Sexuality.* Clairview Books 2014.
2 Michel Odent. *The Scientification of Love.* Free Association Books. London 1999.
3 Michel Odent. *The Functions of the Orgasms: the Highways to Transcendence.* Pinter and Martin. London 2009.

Chapter 2

1 West RG. 'Relative Land-Sea-Level Changes in Southeastern England during the Pleistocene.' *Phil. Trans. R. Soc.* Lond. 1972; 272: 87-98.
2 Williams AN, Ulm S, Sapienza T, Lewis S, Turney CS. 'Sea-level change and demography during the last glacial termination and early Holocene across the Australian continent.' *Quat. Sci. Rev.* 2018;182: 144–54.
3 Benjamin J, O'Leary M, McDonald J, Wiseman C, McCarthy J, Beckett E, et al. (2020) 'Aboriginal artefacts on the continental shelf reveal ancient drowned cultural landscapes in northwest Australia.' *PLoS ONE* 15(7): e0233912. https://doi.org/10.1371/journal.pone.0233912
4 Bailey GN, Harff J, Sakellariou D, editors. *Under the Sea: Archaeology and Palaeolandscapes of the Continental Shelf.* Cham: Springer; 2017.
5 Bailey G, Jöns H. 'The Baltic and Scandinavia: Introduction.' In: Bailey G, Galanidou N, Peeters H, Jöns H, Mennenga M, editors. *The Archaeology of Europe's Drowned Landscapes.* Cham: Springer; 2020, April 10, pp. 27–38. Available from https://link.springer.com/chapter/10.1007/978-3-030-37367-2_2

6 Tierney JE, Zhu J, et al. 'Glacial cooling and climate sensitivity revisited.' *Nature* 2020;584: 569-573

7 https://norwegianscitechnews.com/wp-content/uploads/2017/07/chimp-eve-allan-krill-ntnu-130717.pdf

Chapter 3

1 Melis, Paolo (2002). 'Un Approdo della costa di Castelsardo, fra età nuragica e romana.' In: *L'Africa romana: atti del 14.* Convegno di studio, 7-10 dicembre 2000, Sassari, Italia. Roma, Carocci editore. V. 2, p. 1331-1343: ill. (Collana del Dipartimento di Storia dell'Università degli Studi di Sassari. N. S., 13.2; Pubblicazioni del Centro di studi interdisciplinari sulle Province romane dell'Università degli studi di Sassari, 13.2). ISBN 88-430-2429-9. Conference or Workshop Item.

2 https://www.researchgate.net/publication/215481837_The_human_colonization_of_Sardinia_a_Late-Pleistocene_human_fossil_from_Corbeddu_cave

3 Yirka, Bob (1 March 2012). 'Evidence suggests Neanderthals took to boats before modern humans.' phys.org. Retrieved 5 May 2016.

4 Marshall, Michael (29 February 2012). 'Neanderthals were very ancient mariners.' *New Scientist*. Retrieved 5 May 2016.

5 Charles Q. (15 November 2012). 'Ancient Mariners: did Neanderthals sail to Mediterranean islands?' *LiveScience*. Retrieved 5 May 2016.

6 Ferentinos G, Gkioni M, et al. 'Early seafaring activity in the southern Ionian Islands, Mediterranean Sea.' *Journal of Archaeological Science*, 2012; 39: 2167-2176.

7 Stringer CB, Finlayson JC, et al. 'Neanderthal exploitation of marine mammals in Gibraltar.' *Proc. Natl. Acad. Sci. USA* 2008 Sep. 23;105(38):14319-24. doi: 10.1073/pnas.0805474105. Epub 2008 Sep. 22.

8 Bruce Howe. 'The Palaeolithic of Tangier, Morocco: excavations at Cape Ashakar, 1939-1947.' American School of Prehistoric Research. Peabody Museum. Harvard University. 1967. Bulletin No 22.

9 Ennouchi, Émile (1962). 'Un neandertalien: L'Homme du Jebel Irhoud (Maroc).' *Anthropologie* (66): 279–299.

10 Ennouchi, Émile (1962). 'Un crâne d'Homme ancien au Jebel Irhoud (Maroc).' *Comptes Rendus de l'Académie des Sciences* (254): 4330–4332.

11 Hublin JJ, Ben-Ncer A, et al. 'New fossils from Jebel Irhoud, Morocco and the pan-African origin of Homo sapiens.' *Nature* 2017 Jun. 7;546(7657):289-292. doi: 10.1038/nature22336

12 Richter D, Grun R, et al. 'The age of the hominin fossils from Jebel Irhoud, Morocco, and the origins of the Middle Stone Age.' *Nature* 2017 Jun. 7;546(7657):293-296. doi: 10.1038/nature22335

13 Ewen Callaway. 'Oldest *Homo sapiens* fossil claim rewrites our species' history.' https://www.nature.com/news/oldest-homo-sapiens-fossil-claim-rewrites-our-species-history-1.22114.

14 Morwood MJ, Van Oosterzee P. *A new human: the discovery of the Hobbits of Flores.* Washington DC: Smithsonian Books 20.

15 Bird MI, Condie SA, O'Connor S, et al. 'Early human settlement of Sahul was not an accident.' *Scientific Reports* 2019; 9: 8220.

16 Bradshaw CJA, Ulm S, Williams AN, et al. 'Minimum founding populations for the first peopling of Sahul.' *Nature Ecology & Evolution* 2019; 3: 1057–1063.

17 O'Connor S, et al. 'Pelagic Fishing at 42,000 Years Before the Present and the Maritime Skills of Modern Humans.' *Science* 2011; 334(6059):1117-21 DOI: 10.1126/science.1207703.

18 Dennis Normile. 'Explorers successfully voyage to Japan in primitive boat in bid to unlock an ancient mystery.' *Science* Jul. 10, 2019. https://www.sciencemag.org/news/2019/07/explorers-voyage-japan-primitive-boat-hopes-unlocking-ancient-mystery

19 Moreno-Mayar JV, Vinner L, et al. 'Early human dispersals within the Americas.' *Science* 2018 Nov. 8. Doi: 10.1126/science.aav2621.

20 Posth C, Nakatsuka N, et al. 'Reconstructing the deep population history of Central and South America.' *Cell* 2018 Nov. 8. Doi. org/10.1016/j.cell. 2018.10.027.

21 Erlandson JM, Graham MH, et al. 'The Kelp Highway Hypothesis: Marine Ecology, the Coastal Migration Theory, and the Peopling of the Americas.' *The Journal of Island and Coastal Archeology.* 2007:2;161-174.

22 Holen SR, Deméré TA, Fisher DC, et al. 'A 130,000-year-old archaeological site in southern California, USA.' *Nature* April 2017; 544: 479-483. doi:10.1038/nature22065.

23 Ferraro JV, Binetti LA, et al. 'Contesting early archaeology in California.' *Nature* 2018;554: 479-483.

24 Ruth Gruhn. 'Observations Concerning the Cerutti Mastodon Site.' *PaleoAmerica* 2018; 4 (2): 101-102. https://doi.org/10.1080/20555 563.2018.1467192.

25 Pontus Skoglund, Swapan Mallick et al. 'Genetic evidence for two founding populations of the Americas.' *Nature* 2015;525(7567)104-8. doi: 10.1038/nature14895.

26 Dillehay T, Ocampo C. 'New Archaeological evidence for an early human presence at Monte Verde, Chile.' *Plos One* 2015. *e0141923*. Bibcode:2015 PLoSO..1041923D. doi:10.1371/journal. pone.0141923. PMC 4651426. PMID 26580202.

27 James Hornell. 'The Role of Birds in Early Navigation.' *Antiquity* September 1946; 20 (70): 142-149. DOI: http://dx.doi.org/10.1017/ S0003598X0001953031

28 Jones CM, Braithewaite VA, Healy SD. 'The evolution of sex differences in spatial ability.' *Behav. Neurosci.* 2003 Jun.;117(3):403-11.

29 Clint EK, Sober E, et al. 'Male superiority in spatial navigation: adaptation or side effect?' *Q REV Biol* 2012 Dec.;87(4):289-313.

Chapter 4

1 Prüfer, Kay (2013). 'The complete genome sequence of a Neanderthal from the Altai Mountains.' *Nature*. 505 (1): 43-49. Bibcode:2014 Nature.505...43P. doi:10.1038/nature12886. PMC 4031459. PMID 24352235.

2 Jobling; Hollox; Hurles; Kivisild; Tyler-Smith. 'Human Evolutionary Genetics.' 2nd edition. *Garland Science* 2014.

3 Jacobs, G. S., Hudjashov G., Saag L. 'Multiple Deeply Divergent Denisovan ancestry in Papuans.' (2019). *Cell.* 177 (4): 1010-1021.e32. doi:10.1016/j.cell.2019.02.035. ISSN 0092-8674. PMID 30981557.

4 Choudhury, A., Aron, S., Botigué, L.R. et al. 'High-depth African genomes inform human migration and health.' *Nature* 586, 741–748 (2020). https://doi.org/10.1038/s41586-020-2859-7

5 Reich, David; Patterson, Nick; Kircher, Martin; Delfin, Frederick, et al. (2011). 'Denisova Admixture and the First Modern Human

Dispersals into Southeast Asia and Oceania.' *The American Journal of Human Genetics*. 89 (4): 516-20.doi:10.1016/j.ajhg.2011.09.005. PMC 3188841. PMID 21944045.

6 https://theconversation.com/how-the-banjar-people-of-borneo-became-ancestors-of-the-malagasy-and-comorian-people-90476

7 Huffman, TN. 2000. 'Mapungubwe and the origins of the Zimbabwe culture.' South African Archaeological Society Goodwin Series 8: 19-24.

8 https://www.terradaily.com/reports/Neanderthal_fossil_found_in_North_Sea_999.html

9 Hakon Glorstad, Jostein Gundersen, et al. 'Norway: Submerged Stone Age from a Norwegian Perspective. The Archaeology of Europe's Drowned Landscape,' pp 125-140. COASTALRL 2020: volume 35.

10 https://museum.wales/articles/2007-05-11/The-Cave-Men-of-Ice-Age-Wales/

11 Rouse, Greg W.; Goffredi, Shana K.; Johnson, Shannon B.; Vrijenhoek, Robert C. (5 February 2018). 'An inordinate fondness for Osedax (Siboglinidae: Annelida): Fourteen new species of bone worms from California.' *Zootaxa*. 4377 (4): 451. doi:10.11646/zootaxa.4377.4.1

Chapter 5

1 Jessica Johnson & Michel Odent. *We are All Water Babies*. Celestial Arts. London 1995.

2 'Les Dauphins sauvent trois personnes', Réseau Cétacés No 12, Paris, 1994.

3 Cochrane, A. & Allen, K. *Dolphins and Their Power to Heal*, Bloomsbury, London, 1992.

4 Lilly, J., *Communication between Man and Dolphin*, Julian Press, New York, 1978.

5 Dobbs, H., *The Magic of Dolphins*, Lutterworth Press, Guilford, 1984.

6 Dobbs, H., *Journey into Dolphin Dreamtime*, Jonathan Cape, London, 1992.

7　Smith, B., 'Using Dolphins to Elicit Communication from an Autistic Child.' The pet Connection, Minneapolis Center for the Study of Human-Animal Relationships and Environments, 1984.

8　Ewen Callaway. 'Ancient dog DNA reveals 11,000 years of canine evolution.' *Nature* 2020; 587:20.

9　Anders Bergström, et al. 'Origins and genetic legacy of prehistoric dogs.' *Science*, 2020 DOI: 10.1126/science.aba9572

10 Bray EE, Gitaniali E, et al. 'Early-Emerging and Highly-Heritable Sensitivity to Human Communication in Dogs.' BioRxiv. doi: https://dio.org/10.1101/2021.03.17.434752

11 Ostrander EA, Wayne RW, et al. 'Demographic history, selection and functional diversity of the canine genome.' *Nat. Rev. Genet.* 2017 Dec.; 18(12):705-720.

12 Perri, A; Feuerborn TR, et al. 'Dog domestication and the dual dispersal of people and dogs into the Americas.' *PNAS* February 9, 2021. 118 (6) e2010083118; https://doi.org/10.1073/pnas.2010083118

13 da Silva Coelho FA, Gill S, Tomlin CM, Heaton TH, Lindqvist C. 'An early dog from southeast Alaska supports a coastal route for the first dog migration into the Americas.' *C.Proc. Biol. Sci.* 2021 Feb. 24;288(1945):20203103. doi: 10.1098/rspb.2020.3103. Epub 2021 Feb. 24. PMID: 33622130.

Chapter 6

1　Hill, Kyle (14 July 2014). 'How Science Could Make a Chimp Like DAWN OF THE PLANET OF THE APES Caesar.' archive.nerdist.com

2　Monahan M, Boelaert K, Jolly K, et al. 'Costs and benefits of iodine supplementation for pregnant women in a mildly to moderately iodine-deficient population: a modelling analysis.' *Lancet Diabetes Endocrinol* 2015 Aug. 7. pii: S2213-8587(15)00212-0. doi: 10.1016/S2213-8587(15)00212-0. [Epub ahead of print]

3　William F. Perrin; Bernd Würsig; J.G.M. Thewissen (2009). *Encyclopedia of Marine Mammals*. Academic Press. p. 150. ISBN 978-0-08-091993-5.

4 Marino, Lori (31 December 2004). 'Cetacean Brain Evolution: Multiplication Generates Complexity.' *International Journal of Comparative Psychology*. 17 (1): 1–16.

5 https://synapsida.blogspot.com/2012/01/first-dolphin-in-amazon.html

6 Jeffrey Schwartz, Ian Tattersall. *The Human Fossil Record*. Wiley-Blackwell 2003.

7 Amano, H.; Kikuchi, T.; Morita, Y.; Kondo, O.; Suzuki, Hiromasa; et al. (Aug. 2015). 'Virtual Reconstruction of the Neanderthal Amud 1 Cranium' (PDF). *American Journal of Physical Anthropology*. 158 (2): 185–197. doi:10.1002/ajpa.22777. hdl:10261/123419

8 Gary Lynch, Richard Granger. *Big Brain*. St Martin's Griffin. New York 2008.

9 De Catanzaeo D. *Suicide and self-damaging behavior: a sociobiological perspective*. New York: Academic Press, 1981.

Chapter 7

1 Westenhöfer, Max (1942). 'Der Eigenweg des Menschen. Dargestellt auf Grund von vergleichend morphologischen Untersuchungen über die Artbildung und Menschwerdung.' Berlin: Verlag der Medizinischen Welt, W. Mannstaedt & Co. pp. 309–312. OCLC 311692900

2 https://royalsocietypublishing.org/doi/pdf/10.1098/rsbm.1986.0008.

3 Christopher R. Boyer, 'Geographic Regionalism and Natural Diversity', in *A Companion to Mexican History and Culture*, ed. William H. Beezley. Oxford: Wiley-Blackwell 2011, p. 126

4 Carl Sauer. 'Seashore—Primitive Home of Man?' American Philosophical Society 1962b. Proceeding. 106:41-47.

5 McClure RD (October 1935). 'Goitre Prophylaxis with Iodized Salt'. *Science*. 82 (2129): 370–371.

6 Algis Kuliukas. *Elaine Morgan: 100 years Towards Origins*. Kindle edition 2020.

7 Daryl Leeworthy. *Elaine Morgan: A Life Behind the Screen*. Seren Books 2020.

Chapter 8

1 Plourde M, Cunnane SC. 'Extremely limited synthesis of long chain polyunsaturates in adults: implications for their dietary essentiality and use as supplements.' *Appl. Physiol. Nutr. Metab.* 2007; 32(4):619-34.

2 Burdge GC, Wootton SA. 'Conversion of alpha-linolenic acid to eicosapentaenoic, docosapentaenoic and docosahexaenoic acids in young women.' *Br. J. Nutr.*, 88 (2002), pp. 411-420.

3 Burdge GC, Jones AE, Wootton SA. 'Eicosapentaenoic and docosapentaenoic acids are the principal products of a-linolenic acid metabolism in young men.' *British Journal of Nutrition* (2002), 88, 355–363.

4 Stark AH, Reifen R, Crawford MA. 'Past and present insight on alpha-linolenic acid and the omega-3 fatty acid family.' *Crit. Rev. Food Sci. Nutr.* 2016; 56(14): 2261-7.

5 Odent M, McMillan L, Kimmel T. 'Prenatal care and sea fish.' *Eur. J. Obstet. Gynecol.* 1996; 68:49-51.

6 Lesley F Meeson. 'The effect on birth outcomes of discussions in early pregnancy, emphasising the importance of eating fish.' PhD thesis. University of Wolverhampton. July 2007.

7 Ogundipe E, Tusor N, et al. 'Randomized controlled trial of brain specific fatty acid supplementation in pregnant women increases brain volumes on MRI scans of their newborn infants.' *Prostaglandins Leukot. Essent. Fatty Acids* 2018; 138:6-13.

8 Pickens WL, Warner RR, et al. 'Characterization of vernix caseosa: water content, morphology, and elemental analysis.' *Journal of Investigative Dermatology* 2000;115 (5): 875-881.

9 Jenkins DT, Wysocki SJ, Davies DM. 'Amniotic fluid squalene: A useful test in prolonged pregnancy.' *Aust. N. Z. J. Obstet. Gynaecol.* 1982; 22: 135-7.

10 Wang DH, Ran-Ressler R, et al. 'Sea Lions develop Human-like Vernix Caseoa delivering branched fats and squalene to the GI tract.' *Sci. Rep.* 2018; 8(1):7478. doi: 10.1038/s41598-018-25871-1. PMID: 29748625

11 Michel. Odent 'The obstetrical implications of the aquatic ape hypothesis.' In: *Was Man More Aquatic in the Past?* Algis Kuliukas (Editor), Marc Verhaegen (Editor) *Bentham Science* 2011

12 Young SM, Benyshek DC. 'In search of human placentophagy: a cross-cultural survey of human placenta consumption, disposal practices, and cultural beliefs.' *Ecology of Food and Nutrition* 2010; 49: issue 6. Pages 467-484.

13 Odent, Michel. 'Obstetrical implications of waterside hypothesis.' *Journal of Prenatal and Perinatal Psychology & Health 2012*; 26 (3).

14 Kristal MB. 'Enhancement of opioid-mediated analgesia: A solution to the enigma of placentophagia.' *Neuroscience & Biobehavioral Reviews* 1991; 15 (3): 425-435.

15 Takushi Kishida, JGM Thewissen, Takashi Hayakawa, et al. 'Aquatic adaptation and the evolution of smell and taste in whales.' Zoological letter 2015; 1:9. https://doi.org/10.1186/s40851-014-0002-z

16 Davis RW, Pierotti VR, et al. 'Lipoproteins in pinnipeds: analysis of a high molecular weight form of apolipoprotein E.' *Journal of Lipid Research* 1991; 32(6):1013-23.

17 Boule M. *L'Homme fossile de La Chapelle-aux-Saints* [The Fossil Man from La Chapelle-aux-Saints] (Masson, Paris) French

18 Trinkaus E, Samsel M, Villotte S. 'External auditory exostoses among western Eurasian late Middle and Late Pleistocene humans.' *PLoS One*. 2019 Aug 14;14(8): e0220464.

19 Simas V, Hing W, Furness J, et al. 'The Prevalence and Severity of External Auditory Exostosis in Young to Quadragenarian-Aged Warm-Water Surfers.' *Sports* 2020, 8(2), 17; https://doi.org/10.3390/sports8020017

20 Smith-Guzmán NE, Cooke RG. 'Cold-water diving in the tropics? External auditory exostoses among the Pre-Columbian inhabitants of Panama.' *American Journal of Physical Anthropology* 2019 Mar.;168(3):448-458. doi: 10.1002/ajpa.23757. Epub 2018 Dec. 21

21 Rhys-Evans PH, Cameron M. 'Aural exostoses (surfer's ear) provide vital fossil evidence of an aquatic phase in Man's early evolution.' Ann R Coll Surg. Engl. 2017 Nov.; 99(8):594-601. doi: 10.1308/rcsann.2017.0162. Epub 2017 Sep.

22 Pulvers JN, Journiac N, et al. 'MCPH1: a window into brain development and evolution.' *Front. Cell. Neurosci.*, 27 March 2015. https://doi.org/10.3389/fncel.2015.00092

23 Shi Lei, et al. (27 March 2019). 'Transgenic rhesus monkeys carrying the human MCPH1 gene copies show human-like neoteny of brain

development.' *Chinese National Science Review.* 6 (3): 480–493. doi:10.1093/nsr/nwz043. Retrieved 30 December 2019.

Chapter 9

1 http://rsc03.net/Swimming.html

2 https://www.spasbudapest.com/20180917-when-did-humans-start-to-swim

3 McGraw MB. 'Swimming Behavior of the Human Infant.' *Journal of Pediatrics* 1939; 15: 485-90.

4 Jessica Johnson & Michel Odent. 'What newborn babies can do.' In *We are all water babies.* Celestial arts. London 1994

5 Renato Bender, Nicole Bender. 'Brief communication: Swimming and diving behavior in apes (Pan troglodytesandPongo pygmaeus): First documented report.' *American Journal of Physical Anthropology*, 2013; DOI: 10.1002/ajpa.22338

6 Trinkaus E, Samsel M, Villotte S. 'External auditory exostoses among western Eurasian late Middle and Late Pleistocene humans.' *PLoS One.* 2019 Aug. 14;14(8): e0220464.

7 Villa P, Soriano S, et al. 'Neandertals on the beach: Use of marine resources at Grotta dei Moscerini (Latium, Italy).' *Plos One* 2020;15(1): e0226690. doi: 10.1371/journal.pone.0226690. eCollection 2020.

Chapter 10

1 Endevelt-Shapira Y, Shushan S, Roth Y, Sobel N. 'Disinhibition of olfaction: human olfactory performance improves following low levels of alcohol.' *Behav. Brain Res.* 2014 Oct 1;272:66-74. doi: 10.1016/j.bbr.2014.06.024. Epub 2014 Jun 25.

2 Odent M. 'The early expression of the rooting reflex.' Proceedings of the 5th International Congress of Psychosomatic Obstetrics and Gynaecology, Rome 1977. London: Academic Press, 1977: 1117-19.

3 Odent M. *'L'expression précoce du réflexe de fouissement.'* In : *Les cahiers du nouveau-né.* Paris 1978 ; 1-2 : 169-185

4 Marlier L, Schaal B, Soussignan R. 'Orientation responses to bio-logical odours in the human newborn. Initial pattern and postnatal plasticity.' *C R Acad Sci III*. 1997 Dec.;320(12):999-1005. PMID: 9587477 [PubMed—indexed for MEDLINE]

5 Varendi H, Porter RH, Winberg J. 'The effect of labor on olfactory exposure learning within the first postnatal hour.' *Behav. Neurosci.* 2002 Apr;116(2):206-11. PMID: 11996306 [PubMed—indexed for MEDLINE]

6 Cernoch JM, Porter RH. 'Recognition of maternal axillary odors by infants.' *Child Dev.* 1985 Dec.;56(6):1593-8.

7 Sullivan RM, Toubas P. 'Clinical usefulness of maternal odor in new-borns: soothing and feeding preparatory responses.' *Biol. Neonate.* 1998 Dec.;74(6):402-8.

8 Varendi H, Christensson K, Porter RH, Winberg J. 'Soothing effect of amniotic fluid smell in newborn infants.' *Early Hum. Dev.* 1998 Apr. 17;51(1):47-55.

9 Varendi H, Porter RH, Winberg J. 'Attractiveness of amniotic fluid odor: evidence of prenatal olfactory learning?' *Acta Paediatr.* 1996 Oct.;85(10):1223-7.

10 Schaal B, Marlier L, Soussignan R. 'Olfactory function in the human fetus: evidence from selective neonatal responsiveness to the odor of amniotic fluid.' *Behav. Neurosci.* 1998 Dec.;112(6):1438-49.

11 Marlier L, Schaal B, Soussignan R. 'Neonatal responsiveness to the odor of amniotic and lacteal fluids: a test of perinatal chemosensory continuity.' *Child Dev.* 1998 Jun.;69(3):611-23.

12 Marlier L, Schaal B, Soussignan R. 'Bottle-fed neonates prefer an odor experienced in utero to an odor experienced postnatally in the feeding context.' *Dev. Psychobiol.* 1998 Sep.;33(2):133-45.

13 Orefice G, Modafferi N, et al. 'Archaic reflexes in normal elderly peo-ple.' *Acta Neurol.* 1991 Feb.;13(1):19-24.

14 Elkadry E, Kenton K, et al. 'Do mothers remember key events during labor?' *Am. J. Obstet. Gynecol.* 2003;189:195-200.

15 Wang F, Li J, et al. 'The GABA(A) receptor mediates the hypnotic activity of melatonin in rats.' *Pharmacol. Biochem. Behav.* 2003 Feb.;74(3):573-8.

16 Tysio R, Nsardou R, et al. 'Oxytocin-mediated GABA inhibi-tion during delivery attenuates autism pathogenesis in rodent

offspring.' *Science*. 2014 Feb. 7;343(6171):675-9. doi: 10.1126/science.1247190.

17 Cohen M, Roselle D, Chabner B, Schmidt TJ, Lippman M. 'Evidence for a cytoplasmic melatonin receptor.' *Nature*. 1978; 274:894-895.

18 Sharkey, James Thomas, 'Melatonin Regulation of the Oxytocin System in the Pregnant Human Uterus' (2009). Electronic Theses, Treatises and Dissertations. Paper 1791. http://diginole.lib.fsu.edu/etd/1791

19 Olcese J, Beesley S. 'Clinical significance of melatonin receptors in the human myometrium.' *Fertil. Steril.* 2014 Jul. 8. pii: S0015-0282(14)00566-4. doi:10.1016/j.fertnstert.2014.06.020. [Epub ahead of print]

20 Schlabritz-Loutsevitch N, Hellner N, Middendorf R, Muller D, Olcese J. 'The human myometrium as a target for melatonin.' *J. Clin. Endocrinol* Metab. 2003;88(2):908-913.

21 Sharkey JT, Puttaramu R, Word RA, Olcese J. 'Melatonin synergizes with oxytocin to enhance contractility of human myometrial smooth muscle cells.' *J. Clin. Endocrinol Metab.* 2009 Feb.;94(2):421-7. doi: 10.1210/jc.2008-1723. Epub 2008 Nov 11.

22 Bagci S, Berner AL, et al. 'Melatonin concentration in umbilical cord blood depends on mode of delivery.' *Early Human development* 2012; 88(6):369-373.

23 Ferguson, J.K.W. 'A study of the motility of the intact uterus at term.' *Surg. Gynecol. Obstet.* 1941. 73: 359-66.

24 Flint AP, Forsling ML, Mitchell MD, Turnbull AC. 'Temporal relationship between changes in oxytocin and prostaglandin F levels in response to vaginal distension in the pregnant and puerperal ewe.' *J. Reprod. Fertil.* 1975 Jun.;43(3):551-4.

25 Flint AP, Forsling ML, Mitchell MD. 'Blockade of the Ferguson reflex by lumbar epidural anaesthesia in the parturient sheep: effects on oxytocin secretion and uterine venous prostaglandin F levels.' *Horm. Metab. Res.* 1978 Nov.;10(6):545-7.

26 Newton N, Foshee D, Newton M. 'Experimental inhibition of labor through environmental disturbance.' *Obstetrics and Gynecology* 1967; 371-377.

27 Odent M. 'The fetus ejection reflex.' *Birth* 1987; 14: 104-105.

28 Newton N. 'The fetus ejection reflex revisited.' *Birth* 1987; 14: 106-108.

29 Odent M. 'Fear of death during labour.' *Journal of Reproductive and Infant Psychology* 1991; 9: 43-47.

30 Odent M. 'The second stage as a disruption of the fetus ejection reflex.' *Midwifery Today* 2000;55:12.

31 Wulf Schiefenhovel. *Childbirth among the Eipos, New Guinea*. Film presented at the Congress of Ethnomedicine. 1978.Gottingen. Germany

32 https://www.thetimes.co.uk/article/knitting-is-such-a-relief-from-lifes-great-terrors-ng5zcdwxg9h

33 Odent M. 'Knitting midwives for drugless childbirth?' *MidwiferyToday* 2004; vol 71: 21-22.

34 Odent M. 'Knitting needles, cameras and electronic fetal monitors.' *MidwiferyToday*, Spring 1996; 37: 14-15.

Chapter 11

1 Sigmund Freud. *The Future of an Illusion*. Penguin 7 August 2008.

2 Michel Odent. *The Functions of the Orgasm. The Highways to Transcendence*. Pinter & Martin. London 2009.

3 Araujo L, Ribeiro O, et al. 'The Role of Existential Beliefs Within the Relation of Centenarians' Health and Well-Being.' *J. Relig. Health*. 2017 Aug.;56(4):1111-1122. DOI: 10.1007/s10943-016-0297-5

4 Devinsky, J.; Schachter, S. (2009). 'Norman Geschwind's contribution to the understanding of behavioral changes in temporal lobe epilepsy: The February 1974 lecture'. *Epilepsy & Behavior*. 15 (4): 417–24. doi:10.1016/j.yebeh.2009.06.006. PMID 19640791. S2CID 22179745.

5 Zhihao Guo, Baotian Zhao, et al. 'Effective connectivity among the hippocampus, amygdala, and temporal neocortex in epilepsy patients: A cortico-cortical evoked potential study.' *Epilepsy Behav*. 2021 Jan. 8;115:107661. doi: 10.1016/j.yebeh.2020.107661.

6 Madeleine Odent. 'Joie'. In: *Rayons du Soir*. Les Presses de Monteil. Pessac 1978.

Chapter 12

1 Michel Odent. *Genese de l'homme écologique*. Epi. Paris 1979. Second édition Le Hetre Myriadis. Paris 2019.

2 https://www.millenniumassessment.org/documents/document.288.aspx.pdf

3 http://www.unesco.org/new/en/natural-sciences/ioc-oceans/focus-areas/rio-20-ocean/blueprint-for-the-future-we-want/marine-biodiversity/facts-and-figures-on-marine-biodiversity/

4 GK Saba, Burd AB, et al. 'Towards a better understanding of fish-based contribution to ocean carbon flux.' *Limnology and Oceanography.* Feb. 2021. DOI: 10.1016/j.ecss.2021.107255

5 Sala, E., Mayorga, J., Bradley, D. et al. 'Protecting the global ocean for biodiversity, food and climate.' *Nature* (2021). https://doi.org/10.1038/s41586-021-03371-z

6 Caldeira K, Wickett ME. 'Anthropogenic carbon and ocean pH.' *Nature* 2003; 425: 365.

7 Caesar L, McCarthy GD, et al. 'Current Atlantic Meridional Overturning Circulation weakest in last millennium.' *Nature Geoscience* 2021. DOI: 10.1038/s41561-021-00699-z

Chapter 13

1 Whitehouse, David (January 21, 2003). *'Oldest star chart' found.* BBC. Retrieved 2009-09-29.

2 Odent M. 'Birth under water.' *Lancet* 1983;2:1376-77.

3 Odent M. 'Can water immersion stop labor?' *J. Nurse Midwifery* 1997 Sep-Oct.;42(5):414-6.

4 Odent M. 'What I learned from the first hospital birthing pool.' *MidwiferyToday* 2000;54:16.

5 Odent M. 'A landmark in the history of birthing pools.' *Midwifery-Today* 2000;54:17-8, 69.

6 Daniel Everett. *Don't sleep, there are snakes*. Profile Books 2008.

7 Michel Odent. *Water and Sexuality*. Arkana (Penguin) 1990.

Index

G
Galileo, 97-98
genetics, viii, ix, 3, 26, 28, 39, 62, 107-108
Gibraltar, viii, 14-15, 105
globalization, 91
gorilla, 67, 90
Gulf stream, 93-94

H
Hardy, Alister, 48-49, 52-53
history, 2, 6-7,11, 13, 16, 18, 23, 26-27, 39, 49, 52, 58, 61, 65, 77, 90, 97, 99, 106, 109-110, 117
Homo floresiensis, 17, 27
humanity, xi, xii, 34, 58, 89-91, 93, 95

I
inspiration, 1, 86-87
interbreeding, 28, 90
interdisciplinarity, ix, 47-48, 52
iodine, 41, 51, 54, 109
iodized salt, 51, 110
Ionian islands, 13, 105
IQ, 41

J
Japan, 29, 69, 106
joy, xi

K
kelp highway, 38, 87-88
knitting, 81, 116

L
larynx, 60
LED (light emitting diode), 77

M
Madagascar, viii, 22, 28-29

Mayol, Jacques, 35, 101
McGraw, Myrtle, 66-67, 113
Mediterranean basin, vii, 11, 14-16, 26
melatonin, 77, 114-115
mercury pollution, 94
mermaid, 1-2
microplastics, 94-95
Monte Verde, 20, 107
Morgan, Elaine, ix, 46-47, 49-53, 101, 110
Moroccan coast, 14
Mousterian, 14-15
mystery (mysteries), x, 1, 3, 18, 20, 22, 29, 70, 87, 106
myths, 1, 58
nakedness, 57

N
Neanderthal, 3, 13-15, 24-25, 27, 28, 30-31, 43-44, 60, 63, 71, 91, 105, 107-108, 110
neocortical inhibition, xi, 45, 67, 74-75, 77, 81, 83
Newton, Niles, 78-79, 115-116
nose, 60

O
Olduvai Gorge, 24
olfaction, 74, 113
Olympic games, ix, 68-70
orangutans, 67
orders of magnitude, 98
Ortwin Sauer, Carl, 50, 52
overspecialization, 5, 10
oxytocin, 77-78, 81, 91, 114-115

P
Pacific Rim, vii, 11, 16, 28, 70, 98
painting, 2, 87
palaeontology, 5